Schriftenreihe der Professur für Molekulare Lebensmitteltechnologie

Band 9

Institut für Ernährungs- und Lebensmittelwissenschaften
Professur für Molekulare Lebensmitteltechnologie

Effects of cell wall degradation on the polyphenol content and profile of red berries during juice production

Einflüsse des Zellwandabbaus auf Polyphenolgehalt und -profil roter Beerenfrüchte während der Saftherstellung

Dissertation

zur

zur Erlangung des Doktorgrades (Dr. rer. nat.)

der

Mathematisch-Naturwissenschaftlichen Fakultät

der

der Rheinischen Friedrich-Wilhelms-Universität Bonn

vorgelegt von

Lena Rebecca Larsen

aus

Achim

Bonn 2021

Bibliografische Information der Deutschen Nationalbibliothek
Die Deutsche Nationalbibliothek verzeichnet diese Publikation in der Deutschen Nationalbibliografie; detaillierte bibliographische Daten sind im Internet über http://dnb.d-nb.de abrufbar.
1. Aufl. - Göttingen: Cuvillier, 2021
Zugl.: Bonn, Univ., Diss., 2021

Angefertigt mit Genehmigung der Mathematisch-Naturwissenschaftlichen Fakultät der Rheinischen Friedrich-Wilhelms-Universität Bonn

1. Gutachter: Prof. Dr. Andreas Schieber

IEL - Molekulare Lebensmitteltechnologie, Universität Bonn

2. Gutachter: Prof. Dr. Matthias Wüst

IEL - Lebensmittelchemie, Universität Bonn

Tag der Promotion: 22.10.2021

Erscheinungsjahr: 2021

Nonnenstieg 8, 37075 Göttingen
Telefon: 0551-54724-0
Telefax: 0551-54724-21
www.cuvillier.de

1. Auflage, 2021
Gedruckt auf umweltfreundlichem, säurefreiem Papier aus nachhaltiger Forstwirtschaft.

ISBN 978-3-7369-7518-7
eISBN 978-3-7369-6518-8

Ich bin dankbar für die Steine, die mir in den Weg gelegt wurden.

Ohne sie wäre ich nicht über meine Stärken gestolpert.

unbekannt

Table of contents

Preliminary remarks

List of abbreviations

ANOVA	analysis of variance
cya	cyanidin
DA	degree of acetylation
DAD	diode array detector
del	delphinidin
dha	3-Desoxy-D-lyxo-2-heptulosaric acid
DM	degree of methylation
EMP	enzymatically modified pectin
FID	flame ionization detector
galAc	galacturonic acid
GC	gas chromatography
HG	homogalacturonan
HPLC	high-performance liquid chromatography
HPSEC	high-performance size exclusion chromatography
kdo	Ketodeoxyoctonic acid/3-Deoxy-D-manno-2-octulosonic acid
mal	malvidin
MHR	modified hairy regions

Mp	peak molecular weight
MW	molecular weight
pel	perlagonidin
peo	peondin
PG	polygalacturonase
PL	pectin lyase
PME	pectin methylesterase
RG	rhamnogalacturonan
RI	retention index
SBP	sugar beet pectin
TAC	total anthocyanin content
UAEM	ultrasound-assisted enzymatic maceration
UHPLC	ultra high-performance liquid chromatography
UMP	ultrasound modified pectin

List of publications

Weber, F., & Larsen, L. R. (2017). Influence of fruit juice processing on anthocyanin stability. *Food Research International*, *100*, 354-365.

Larsen, L. R., Buerschaper, J., Schieber, A., & Weber, F. (2019). Interactions of anthocyanins with pectin and pectin fragments in model solutions. *Journal of Agricultural and Food Chemistry*, *67*(33), 9344-9353.

Larsen, L. R., van der Weem, J., Caspers-Weiffenbach, R., Schieber, A., & Weber, F. (2021). Effects of ultrasound on the enzymatic degradation of pectin. *Ultrasonics Sonochemistry*, 72, 105465.

Conferences

Larsen, L. R., Buhrandt, A., Schieber, A., & Weber, F. (2019). Steigerung des Pektinabbaus durch Ultraschalleinwirkung auf Enzympräparate während der Saftherstellung. 47. Deutscher Lebensmittelchemikertag in Berlin, Germany, September 17-19, 2018, *Lebensmittelchemie*, 73, 49-80. [Poster]

Larsen, L. R., Buerschaper, J., Schieber, A., & Weber, F. (2018). Complexation of anthocyanins by pectin fragments in model fruit juice. 12th INSAH World Congress on Polyphenols Applications, Bonn, Germany, September 25-28, 2018, *Abstracts Books,* 6 (*3*), 73. [Oral presentation]

Larsen, L. R., Schieber, A., & Weber, F. (2019). Interaktionen von Anthocyanen mit nativem, enzymatisch und Ultraschall-modifiziertem Pektin in Modelllösungen. 48. Deutscher Lebensmittelchemikertag in Dresden, Germany, September 17-19, 2019, *Lebensmittelchemie*, 73, S053-S053. [Poster]

Larsen, L. R., van der Weem, J., Schieber, A., & Weber, F. (2020). Ultrasound-assisted enzymatic maceration enhances pectin degradation. P.T3.031. 34th International EFFoST 2020 in Israel, online conference, November 10-12. [Poster]

Declaration of contribution as co-author

The following co-authors contributed to the papers presented in Chapter 2 and 3:

Prof. Dr. rer. nat. habil. Andreas Schieber contributed to the publications as the supervisor of this thesis and proofread all manuscripts.

PD Dr. rer. nat. habil. Fabian Weber advised on the design of experimental work and supported the interpretation and publication of the results. As the corresponding author, he proofread the manuscripts and handled the formal aspects of the publications.

Annerose Buhrandt assisted with experiments of ultrasound-assisted enzymatic activities.

Julia Buerschaper assisted by modifying of pectin solutions and the anthocyanins complexation experiments. She also contributed to the subsequent UHPLC analysis.

Judith van der Weem contributed to the establishment of the continuous circulation system including ultrasound probe. She conducted the UAEM experiments and contributed to the subsequent pectin analysis.

Rita Caspers-Weiffenbach supported several lab works and particularly the analysis of enzyme activities.

Summary

Red berry juices are a rich source of health-promoting plant phenolic compounds. Among these, especially anthocyanins, are associated with various positive effects on human health caused by their antioxidative properties. Furthermore, owing to their intense color, they contribute to the organoleptic quality of the juices considerably. However, current techniques used in juice production lead to a high loss and a significant alteration in the anthocyanins profile. These changes are caused by an insufficient cell wall degradation, resulting in incomplete extraction and the complexation of the extracted polyphenols with the degraded cell wall polysaccharides. Depending on their solubility, anthocyanin-polysaccharide complexes may be removed from the juice during pressing and are eventually discarded with the press cake. To overcome the incomplete extraction, tailormade enzyme preparations with diverse pectinolytic activities are currently applied, improving cell wall degradation. Nevertheless, berry cell walls consist of a particularly pectin-rich complex polysaccharide network and even optimized enzyme treatments lead to considerable losses of anthocyanins since interactions between this complex polysaccharide network and anthocyanins are extremely diverse. However, the resulting complexes may also bear protective properties, preventing sensitive anthocyanins from degradation.

This dissertation details out novel approaches to solve the aforementioned problems in two interconnected steps: First, with regard to pectin degradation, a gentle process using ultrasound technology was investigated. Next, the binding behavior of the resulting pectic polysaccharide residues toward a broad range of anthocyanins was studied in the second step of this thesis.

Previous research shows a promising potential of ultrasound technology as a gentle technique for juice production. Until now, scientists have focused on higher juice yields as well as an improved extraction of phytochemical compounds such as phenolic acids, flavonoids, and especially anthocyanins. This was shown to be achieved with shortened juice processing time and lower processing temperatures. However, the underlying mechanism of cell wall degradation has remained largely neglected. This thesis illustrates the potential of the combined application of ultrasound and enzymes, referred to as ultrasound-assisted enzymatic maceration (UAEM) to reveal the mechanism that goes on during cell wall degradation. In a continuous system, UAEM achieved a considerably higher total process output, which was proven by the specific product formation of the respective enzymes at reduced temperature (30 °C) compared to the conventional batch maceration at 50 °C. According to literature, the enhanced degradation is assumed to be caused by synergistic effects of ultrasound on enzyme activities. Therefore, in this thesis, the three main pectinolytic activities of enzyme preparations, including polygalacturonase (PG), pectin lyase (PL), and pectin methylesterase (PME), were examined under simultaneous sonication. While the effective enzyme activity of PL could be enhanced significantly by ultrasound treatment, the PME activity was not positively affected. The different impacts can be explained by the enzyme's diverse conformations that are individually affected by ultrasonication. The findings demonstrate that ultrasound treatments, depending on the intensity, can lead to an increase or inactivation of enzyme activities.

The resulting pool of polysaccharides and oligosaccharides after UAEM was characterized by their structural properties, including molecular weight, monosaccharide composition, and substitution pattern (degree of methylation and acetylation). Results reveal that the UAEM treatment at reduced temperatures of 30 °C produced the smallest and simplest polymers in terms of molecular weight and neutral sugar side branches compared to the other exclusive treatments. In particular, the soluble low-molecular-weight fraction (>30 kDa) was enriched in monosaccharides from all pectin subunits. This fraction could bear positive effects on the juice quality as it increases the soluble fiber content and stabilizes polyphenols in liquids.

During red juice production, pectic polysaccharides interact with polyphenols and particularly with anthocyanins by forming complexes. Insoluble complexes are likely separated with the press cake, while soluble complexes can have several beneficial properties for red juices such as the enhancement of color stability and antioxidative potential. So far, binding mechanisms during complex formation are not fully understood and are thought to strongly depend on the structural features of the reacting molecules and surrounding pH. Several multifaceted pectic polymers are generated during fruit juice production, which have not been considered in anthocyanin complexation so far. Correspondingly, this work aimed also to understand the interactions between modified pectin residues, similar to those produced in maceration and individual anthocyanins varying in their aglycone and glycoside compositions. Pectin modifications by enzymatic and ultrasound-assisted treatments generated polymers structurally different in molecular weight, monosaccharide composition, and substitution pattern (degree of methylation and acetylation). Results indicate that the structure of both pectin and anthocyanin influence these complexations, whereas the former predominantly governs the complexation abilities. Linear polymers complexed most (up to 70%) of the total anthocyanins as insoluble complexes after two weeks of storage. Probably, the linear polysaccharides favored the stacking of self-associated anthocyanins. In contrast, the enzymatically modified oligosaccharides formed soluble anthocyanin complexes protecting the sensitive molecules against oxidative degradation during two weeks of storage, as determined by a 10% higher total anthocyanin concentration compared to the pectin-free control. According to these results, hydrophobic interactions between the anthocyanins and galacturonic acid-rich pectin units also seem to play a considerable role besides the frequently discussed ionic and hydrophilic interactions.

In conclusion, the basic research of this thesis contributes to a better understanding of the pectin degradation mechanisms by enzymatic, ultrasound, and UAEM into soluble polysaccharides, which are very likely to complex with polyphenols like anthocyanins during juice production and storage. Furthermore, the study and characterization of anthocyanin-pectin binding mechanisms play an important role in devising strategies that may help in reducing the formation of disadvantageous insoluble complexes and achieve an increase in soluble polyphenol stabilizing complexes in the juice.

Zusammenfassung

Buntsäfte aus roten Beeren zeichnen sich durch besonders hohe Gehalte an gesundheitsfördernden phenolischen Pflanzeninhaltsstoffen aus. Im Vordergrund stehen die antioxidativ wirksamen Anthocyane, die mit diversen positiven Effekten für die menschliche Gesundheit assoziiert werden und mit ihrer intensiven Farbeigenschaft wesentlich zur organoleptischen Qualität der Säfte beitragen. Die derzeit übliche Saftherstellung führt allerdings zu teils hohen Verlusten und zu einer starken Profiländerung der Anthocyane. Die Hauptursachen hierfür liegen einerseits an einem unzureichenden Zellwandabbau und einer damit einhergehenden unvollständigen Extraktion, andererseits in den Interaktionen der extrahierten Polyphenole mit den degradierten Zellwandpolysacchariden. Zwar schützen die resultierenden Komplexe die sensitiven Anthocyane vor Abbaureaktionen, jedoch werden diese in unlöslichen Komplexen derzeit größtenteils im Presskuchen bei der Saftherstellung abgetrennt. Um dem Problem der unvollständigen Extraktion zu begegnen, werden speziell entwickelte Enzympräparate mit diversen pektinolytischen Aktivitäten verwendet, die zu einem verbesserten Zellwandabbau während der gegenwärtigen Maischeenzymierung führen. Da die Zellwand bei Beeren aus einem besonders pektinreichen komplexen Polysaccharid-Netzwerk besteht, ermöglicht dieser Enzymeinsatz eine höhere Saft- sowie Polyphenolausbeute. Nichtsdestotrotz gehen erhebliche Mengen der wertvollen phenolischen Verbindungen in den Pressrückständen verloren.

Die vorliegende Dissertation trägt zur Lösung dieser Probleme in zwei aufeinander aufbauenden Schritten bei: Zunächst erfolgte die Untersuchung eines schonenden Verfahrens unter Verwendung der Ultraschalltechnologie in Bezug auf den pektischen Polysaccharidabbau. Anschließend wurden die resultierenden pektischen Polysaccharid-fragmente auf ihr Bindungsverhalten gegenüber einem breiten Spektrum an Anthocyanen untersucht.

Erste Forschungsergebnisse zeigten, dass die Ultraschalltechnik ein vielversprechendes Potenzial für eine schonende Saftproduktion besitzt. Bisher wurde von höheren

Saftausbeuten sowie verbesserter Extraktion von Verbindungen wie Phenolsäuren, Flavonoiden und insbesondere Anthocyanen unter verkürzten Prozesszeiten und niedriger Temperatur berichtet. Der Mechanismus des Zellwandabbaus blieb hierbei weitgehend unbeachtet. Die vorliegende Arbeit verdeutlicht das Potential der kombinierten Anwendung von Ultraschall und Enzym, der sogenannten ultraschall-gestützten Maischeenzymierung (*ultrasound-assisted enzymatic maceration,* UAEM), anhand des für die Buntsaftherstellung essentiellen Pektinabbaus. In einem kontinuierlichen System konnte mit Hilfe der UAEM bei reduzierter Temperatur (30 °C) eine deutlich höhere Gesamtprozessleistung, gemessen an der spezifischen Produktbildung der jeweiligen Enzyme, erzielt werden als in einer konventionell durchgeführten Mazeration bei 50 °C. Für den verbesserten Polysaccharidabbau werden synergistische Effekte zwischen Ultraschalleffekten und dem enzymatischen Abbau vermutet, welche bisher nur anhand der Polygalacturonase (PG) untersucht wurden. In der vorliegenden Arbeit wurden daher die drei wichtigsten pektinolytischen Aktivitäten der Enzympräparate unter Ultraschallbehandlung untersucht, d.h. neben der PG auch die Pektinlyase (PL) und die Pektinmethylesterase (PME). Während die Enzymaktivität der PL durch die Ultraschallbehandlung gesteigert werden konnte, wurde die PME nicht positiv beeinflusst. Die unterschiedlichen Ergebnisse lassen sich durch die verschiedenen Konformationen der Enzyme erklären, die durch die Ultraschall-behandlung individuell beeinflusst werden. Die Ergebnisse verdeutlichen, dass Ultraschallbehandlungen entweder zu einer Steigerung oder Inaktivierung der Enzymaktivitäten führen, die nicht nur von der Ultraschallintensität abhängig sind, sondern auch von den spezifischen Eigenschaften des Enzyms.

Der entstehende Pool von pektischen Polysacchariden und Oligosacchariden wurde anhand der strukturellen Eigenschaften – einschließlich Molekulargewicht, Mono-saccharidzusammensetzung und Substitutionsmustern (Methylierungs- und Acetylierungsgrad) – der Polymere charakterisiert. Die Ergebnisse zeigen, dass die UAEM-Behandlung bei reduzierten Temperaturen von 30 °C die kleinsten und einfachsten Polymere in Bezug auf das Molekulargewicht und die Neutral-zuckerseitenketten im Vergleich zu den anderen exklusiven Behandlungen erzeugen. Insbesondere wurde die lösliche niedermolekulare Fraktion (>30 kDa) mit spezifischen

Monosacchariden aus allen Pektinsubeinheiten angereichert. Diese Fraktion wirkt sich positiv auf die Saftqualität aus, da sie den löslichen Ballaststoffanteil erhöht und die Polyphenole in Lösungen stabilisiert.

Während der Buntsaftherstellung können die pektischen Polysaccharide mit Polyphenolen und insbesondere den Anthocyanen Komplexe bilden. Sind diese unlöslich, werden sie mit dem Presskuchen abgetrennt, während lösliche Komplexe diverse vorteilhafte Eigenschaften besitzen, wie zum Beispiel die Erhöhung der Farbstabilität und des antioxidativen Potentials. Bisher sind die Bindungsmechanismen der Komplexbildungen noch nicht vollumfänglich geklärt. Sie hängen unter anderem stark von den strukturellen Eigenschaften der reagierenden Moleküle und dem pH-Wert ab. Bei der Fruchtsaftherstellung entstehen diverse strukturell unterschiedliche Polysaccharide, die bei der Komplexierung von Anthocyanen bisher nicht berücksichtigt wurden. Entsprechend war es ein weiteres Ziel dieser Arbeit, die Bindungsaffinitäten zwischen modifiziertem Pektin, wie es während der Maische bei der Saftherstellung entsteht, und diversen Anthocyanen, die sich in ihrer Aglykon- und Glykosid-zusammensetzung unterscheiden, genauer zu untersuchen. Das Pektin wurde enzymatisch und ultraschallgestützt modifiziert, so dass sich die resultierenden Polymere deutlich in ihrer Struktur hinsichtlich des Molekulargewichts, der Monosaccharid-zusammensetzung und der Substitutionsmuster (Methylierungs- und Acetylierungsgrad) unterschieden. Die Ergebnisse verdeutlichen, dass sowohl die Struktur des Pektins als auch die des Anthocyans die Komplexierung beeinflussen, wobei erstere die gewichtigere Rolle spielen. Lineare Polymere komplexierten den Großteil (bis zu 70 %) der gesamten Anthocyane nach einer zweiwöchigen Inkubationszeit in unlöslichen Komplexen. Aus den Ergebnissen kann abgeleitet werden, dass die linearen Polysaccharide die Stapelbildung von selbstassoziierten Anthocyanen begünstigen. Im Gegensatz dazu bilden die enzymatisch modifizierten Oligosaccharide lösliche Anthocyan-Komplexe, welche die sensitiven Moleküle während der zweiwöchigen Lagerung vor oxidativem Abbau schützen (gezeigt anhand einer 10 % höheren Gesamtanthocyankonzentration im Vergleich zur pektinfreien Kontrolle). Den Ergebnissen zufolge spielen neben den häufig diskutierten ionischen und hydrophilen Wechselwirkungen zwischen den Anthocyanen und galakturonsäurereichen Pektineinheiten auch hydrophobe Wechselwirkungen eine wichtige Rolle.

Schlussfolgernd bleibt festzuhalten, dass die Grundlagenforschung dieser Arbeit zu einem besseren Verständnis der Mechanismen des Pektinabbaus durch Enzyme, Ultraschall und UAEM in lösliche Polysaccharide beiträgt, die sehr wahrscheinlich Polyphenole wie Anthocyane während der Saftproduktion und -lagerung komplexieren. Die Charakterisierung von Anthocyan-Pektin-Komplexen liefert einen essentiellen Ansatz, um verlustbringende unlösliche Komplexe zu reduzieren und eine Zunahme von löslichen stabilisierenden Polyphenolkomplexen im Saft zu erreichen.

Chapter 1

General Introduction

1 Red juice as functional foods

We live in an age of significant technological changes, with tangible improvements in every part of life, such as in the food segment. Today, food products are high in quality and microbial safety, rich in aroma and color, and have long shelf lives. Lately, the nutritional value of food has shifted in the focus of the consumers' interests. Nowadays, food should not only be merely satiating, but also has additional functional features. This has inspired academia and the food industry to research for foods with beneficial nutritional values and preferably mild production technologies.

Red berries and their derived juices have gained great popularity due to the growing evidence of health-promoting potential and their appealing taste and color. Scientists are primarily holding the high contents of the phenolic flavonoids, in particular, the red to purple colored anthocyanins responsible for these beneficial qualities.

Besides red berries and their derived products, flavonoids are abundantly spread throughout foods consumed by humans such as fruits, vegetables, nuts, seeds, herbs, spices, and whole grains. Similarly, manufactured products originating from plant sources contain high amounts of polyphenols, e.g., a glass of juice, a cup of coffee, or tee includes about 100 mg polyphenols (Scalbert et al. 2005; Spencer et al. 2008). Flavonoids are synthesized by all vascular plants, as they are responsible for essential functions like natural attractants, repellents, or protection against UV light since they often possess colored and aromatic properties (Pandey and Rizvi 2009). With approximately 10,000 different compounds identified in plants so far, flavonoids represent the most prevalent phenolic groups, while the list is constantly growing (Birt and Jeffery 2013; Cheynier et al. 2013).

In the last decade, a growing number of epidemiological studies and associated meta-analyses concluded that a polyphenol-enriched long-term diet goes along with protection against the development of cancer, cardiovascular diseases, diabetes, osteoporosis, and neurodegenerative diseases. The potential health benefits seem to be countless; in fact, many positive effects are still under investigation and need to be proven in clinical studies (Pandey and Rizvi 2009). Until now, anti-cancerogenic, anti-inflammatory, and immunomodulating properties have been postulated (Arts and Hollman 2005). The flavonoid intake from berries, in particular, demonstrated positive effects against diabetes and slowing of neurodegenerative diseases (Devore et al. 2012).

The health-related effects of polyphenols assessed by their bioaccessibility, bioavailability, and bioactivity not only depend on their structure and quantity but also on their intermolecular interactions with other food matrix compounds. Jakobek (2015) emphasized the significant role of dietary fiber composed of plant cell wall polysaccharides by directly modulating the bioaccessibility and bioavailability of polyphenols and indirectly influencing their nutritional and potential health benefits.

The interaction between polyphenols and plant cell wall polysaccharides occurs spontaneously after the disruption of plant tissue; thus, the naturally separated compounds get into contact. This process happens either during consumption, starting by simple chewing, and continues inside the human digestive tract or occurs under the food processing conditions of several common foods such as fruit juice, purees, jams, or wine.

The complexation is assumed to stabilize polyphenols against degradation processes, which bears two advantages. From a nutritional point of view, complexes are supposed to reach the colonic tract since fibers are only metabolized by the microbiota of the large intestine. Here, they provide a beneficial prebiotic effect and an antioxidant environment towards microbiota in the gastrointestinal tract and an antimicrobial activity against pathogenic bacteria (Molan et al. 2009; Puupponen-Pimiä et al. 2005). From a food technology perspective, complexes enhance the stability of food products like juices regarding taste and color since these are strongly determined by polyphenols. In general, polyphenols significantly contribute to the bitterness, astringency, color, and oxidative stability in food.

Thus, there is a growing awareness from manufacturers' and academia's perspectives of the importance of polyphenol-polysaccharide complexes providing bio- and techno-functional properties, which are particularly important in the red juice sector (Liu et al. 2020; Phan et al. 2017).

This dissertation reflects upon polyphenol-polysaccharide complexation during red juice production, taking a closer look at the cell wall degradation during the maceration process and the corresponding complexation of pectin and anthocyanins, the two main representatives of plants cell wall polysaccharides and polyphenols in red berries. Several novel technologies are part of recent research aiming at gentle process conditions to produce high-quality products rich in natural components and with less undesired process-induced alterations compared to common procedures. As a promising technology, ultrasonication is discussed in the present dissertation to sustainably improve fruit juice production (**Section 4**).

2 Red berry fruits

Red berries providing high amounts of flavonoids are growing in popularity due to their appealing color and sweet taste accompanied by promising health-promoting properties. Berries are consumed fresh or processed into juices, concentrates, purees or jams. Juices (concentrates) and frozen berries are used as food ingredients and natural colorants.

Botanically, a berry is a simple fleshy indehiscent fruit derived from a single ovary of an individual flower. A berry usually contains more than one seed, has a pulpy endocarp and a thin skin or exocarp. Typical red berries are cranberry (*Vaccinium macrocarpon* L.), bilberry (*Viccinium myrtillus* L.), and black currant (*Ribes nigrum* L.). Botanical berries also include tomatoes, eggplants, and guavas. In common parlance, several small, round, and fleshy fruits are also termed berries, e.g., elderberry (*Sambucus nigra* L.), chokeberry (*Aronia melanocarpa* Michx.), raspberry (*Rubus idaeus* L.), and blackberry (*Rubus* sect. *Rubus*). However, in the botanical sense, they mostly belong to stone fruits or aggregate fruits consisting of several smaller fruits. Due to their transport sensitivity and short shelf life, these berries are also defined as soft fruits in food traits.

Berries appear in attractive orange, red, violet, purple, and blue colors like many flowers, fruits, and vegetables. This is the result of the presence of anthocyanins, which account for one of the six flavonoid subclasses of the polyphenol family. These water-soluble pigments, along with other polyphenols, are enclosed in the vacuoles mainly of the deeply colored skin, encapsulated by tonoplast and cytoplasmic lipid membranes of the plant cell wall (Padayachee et al. 2012).

The highest content in total anthocyanin concentration among berries was found in chokeberries (1480 mg/100 g fresh weight) and elderberries (1374.4 mg/100 mg fresh weight) (Wu and Prior 2005). Each berry provides a specific anthocyanin profile that may be consulted for authentication and chemotaxonomic investigations, like a fingerprint pattern. However, it needs to be considered that the anthocyanin profile may be altered depending on the plant's environmental conditions, the processing of the fruit, and the extraction method (Heffels et al. 2015; 2017). Additionally, anthocyanin type and structure show variable stability, which will be introduced in more detail below.

Next to anthocyanins, berries are a rich source of other phenolic compounds like phenolic acids and flavanols, which also bear high antioxidant activities and impart health properties. Well-known are the proanthocyanidins, also referred to as condensed tannins, consisting of polymerized derivatives of flavanols. These are responsible for the astringent taste that occurs due to the complexation of human saliva proteins.

Besides phenolic compounds, berries are a rich source of nutritive compounds like sugars, vitamins, and minerals, all entrapped in the dietary fiber of the cell wall matrix consisting of cellulose, hemicellulose, pectin, lignin, and cutin-like polymers. Soluble solids mainly include sugars with a high proportion of fructose, providing them as valuable for individuals with diabetes diseases.

The total fiber content of berries is relatively high (bilberries 4.9 g/100 g, blackberries 3.2 g/100 g, blackcurrants 3.5 g/100 g, raspberries 4.7 g/100 g) compared to other fruits like apples (2.3 g/100 g), and bananas (2.0 g/100 g) (GMF 2002). Pectin, as the main part of the soluble fibers, undergoes nearly complete fermentation by colonic microflora and leads to the generation of short-chain fatty acids, which lower the colonic pH and prevent colon cancer (Moore et al. 1998). It also has the potential to lower blood cholesterol levels and affect glucose metabolism. Many studies have suggested potential beneficial effects on cholesterolemia and diabetes, the risk for diverticulosis, colon cancer, and coronary heart disease (Voragen et al. 2009). Insoluble fibers like cellulose are not fermented but are essential for intestinal transition activities.

In summary, red berry fruits have been classified as superfoods due to the amount of health-promoting compounds they contain, especially anthocyanins. However, during fruit processing and consumption, these compounds are degraded and affected by oxidation and complexation reactions, as discussed in **Section 3**.

2.1 Anthocyanins

Anthocyanins naturally occur in their glycoside form, where the aglycon, called anthocyanidin, is linked to a sugar moiety. Until now, >500 different anthocyanins and 23 different anthocyanidins have been identified. The six most common anthocyanidins are listed in **Table 1-1** in their 3-glucoside form. Their distribution in fruits and vegetables was examined as: cyanidin 30%, delphinidin 22%, pelargonidin 18%, peonidin 7.5%, malvidin 7.5%, and petunidin 5% (Andersen and Jordheim 2005).

The main structural differences are the number and position of hydroxy and methoxy groups in the B-ring, the nature and number of linked sugars, and the aliphatic and aromatic acids attached to the sugar moieties. Glycosylation increases their water solubility. The most common monosaccharides attached primarily at C3 are glucose, rhamnose, galactose, arabinose, and xylose (in the same order of frequency as mentioned). The structure and substitution variety results in broad shades of colors for anthocyanins due to variances in the absorption intensity (hyperchromic effect) and its wavelength (bathochromic/hypsochromic shifts).

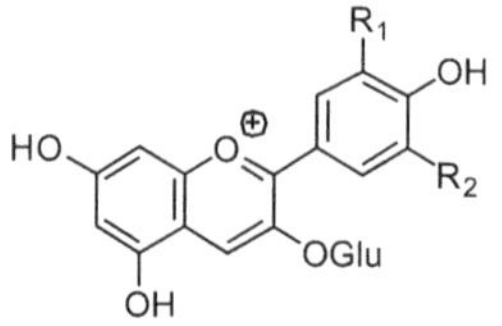

Table 1-1: Chemical structure of anthocyanins

Anthocyanin	R_1	R_2
Perlagonidin-3-glucoside	H	H
Cyanidin-3-glucoside	OH	H
Delphinidin-3-glucoside	OH	OH
Peonidin-3-glucoside	OCH_3	H
Petunidin-3-glucoside	OCH_3	OH
Malvidin-3-glucoside	OCH_3	OCH3

The structure and color of anthocyanins depend on a pH-dependent equilibrium reaction. In acidic milieu (pH <2), the red flavylium cation is predominant, occurring primarily with a chloride counterion. With increasing pH, deprotonation occurs in two steps (pK_a 4 and 7) and the equilibrium shifts via the purple, neutral quinonoid base to the bluish, anionic quinonoid base. Since the quinoid species are thermodynamically unstable, hydration of the flavylium cation at C2 is more likely. Above pH 2, hydration results in the formation of the colorless hemiketal that is prone to tautomerization into isomeric chalcones at pH 6-7. Above pH 7, the chalcone irreversibly degrades by a C-ring opening reaction depending on its substituent groups (Castaneda-Ovando et al. 2009; Trouillas et al. 2016). Red berries contain an acidic environment around pH 3-4, where anthocyanins appear in an equilibrium of the charged flavylium cations, the neutral quinonoid base, and the hemiketal form.

Besides the pH, the individual anthocyanin structure determines the molecule's reactivity and susceptibility toward hydration and oxidative reactions. In general, glycosylated anthocyanins have been shown to be more stable than their aglycone. Anthocyanidins tend to form the chalcone structure eventually followed by an irreversible cleavage, resulting in an aldehyde and phenolic acid. Hydroxy groups and methoxy groups on the B-ring also affect reactions differently. Thus, delphinidin, petunidin, and cyanidin, which contain an *ortho*-di-hydroxylated B-ring, are generally more prone to oxidation than malvidin and peonidin, which lack this pattern (Rossi et al. 2003). Also, the type of glycosylation has an influence on the stability; while pentosides have been demonstrated to be less stable than hexosides, diglucosides were more stable than monoglucosides (Rein and Heinonen 2004). Additionally, the acylation of the glycoside residues by aliphatic or aromatic acids increases the chemical stability and can prolong the half-time values toward thermal degradations (Zhao et al. 2017).

Next to their structure, several exogenous aspects determine anthocyanin stability. Due to their electron-rich aromatic structure, polyphenols are generally strong nucleophiles reacting with other molecules. They also tend to self-associate or are readily oxidized. Polymerization or degradation of the molecule can lead to a decrease in anthocyanin concentration and a reduction in the attractive color. In general, reactions are accelerated tremendously by heat and exposure to light, oxygen, or oxidizing enzymes, all of which are crucial during red juice production.

On the other hand, complex formation like co-pigmentation, including self-association and metal complexation, provide stabilizing effects by protection against hydrolysis and deglycosylation. For example, acylation of hydroxycinnamic acid promotes their stability around neutral pH values due to intramolecular co-pigmentation. This is based on a stacking phenomenon of the hydrophobic acyl moiety and the flavylium nucleus (Castaneda-Ovando et al. 2009; Clifford 2000; Giusti and Wrolstad 2003; Rein 2005; Shikov et al. 2008). A wide range of mostly colorless natural compounds bear a π-electron-enriched system and have been discovered as intermolecular co-pigments, like colorless phenols such as flavonoids and phenolic acids, or alkaloids, amino acids, and organic acids. Co-pigmentation was suggested to be the main mechanism stabilizing the flavylium cation chromophore and thus the color in plants (Mazza and Brouillard 1987; Patras et al. 2010; Rein 2005). However, polymerization to high-molecular-weight adducts can also lead to the formation of brown polymeric pigments, being less attractive in food processing purposes. In general, inter- and intramolecular co-pigmentation with other moieties, polyglycosylated, and polyacylated anthocyanins, were proven to provide greater stability toward changes in pH, heat, and light.

The fruit matrix contains additional putative complexation compounds like polysaccharides and proteins, which may protect anthocyanins from degradation. Even though ample research has been conducted to examine these complexations, detailed binding mechanisms, because of their complexity, still remain under-investigated (Jakobek 2015; Liu et al. 2020; McManus et al. 1985; Phan et al. 2017). The present dissertation focuses on the polysaccharide complexations during red berry juice production.

2.2 Plant cell wall polysaccharides

Pectin, hemicellulose, and cellulose are the main polysaccharides of the primary plant cell wall of fruits and vegetables, with cellulose and hemicellulose providing rigid strength and pectin providing flexibility and fluidity through a gelatinous matrix. Pectin is also the predominant compound in the middle lamella encasing every cell (Chanliaud and Gidley 1999; Ochoa-Villarreal et al. 2012).

The cell wall function includes structural integrity and maintaining the internal pressure. It also has a crucial role in plant growth, cell differentiation, water transport, intercellular communication, and plant defense (Cosgrove 2005; McCann et al. 1990; Ochoa-Villarreal et al. 2012).

The structure of the plant cell wall network is very dynamic and complex; microfibrils of cellulose (30%) are embedded in an amorphous matrix of pectin (35%), while hemicellulose (30%) is stabilized by minor components of protein (glycoproteins, <5%) and phenolic compounds (lignin, insoluble proanthocyanidins, <5%) (data on dry weight basis from Ochoa-Villarreal et al. 2012). Due to its anionic nature, pectin is mainly involved in the regulation of ion transport. It also determines the porosity of walls and controls permeability for enzymes by forming a network independent from the hemicellulose/cellulose network (McCann et al. 1990).

Cellulose microfibrils (Ø approx. 3 nm) are formed by aggregated linear unbranched chains of β-(1,4)-D-glucose residues. These polymers are highly crystalline and insoluble, chemically stable, and resistant to most enzymatic activities.

Hemicelluloses are low molecular weight polysaccharides directly attached to lignin and cellulose. They consist of a heterogeneous group of polysaccharides with a similar structure of β-(1,4)-linked backbones with an equatorial configuration at C1 and C4 (Scheller and Ulvskov 2010). Xyloglucan is the most abundant hemicellulose comprising of a β-(1,4)-glucan backbone, where three out of four glucose residues are substituted with α-xylose at C6 (Ochoa-Villarreal et al. 2012). Other substitutions of galactose, galactosyl-fucose, or arabinose were detected in olives (Vierhuis et al. 2001). Three domains of xyloglucan can be differentiated; the first is enzyme accessible occurring in free loops and cross-links, the second is connected to cellulose microfibrils by hydrogen bonds and extractable by concentrated alkali, and the third is only accessible when cellulose is degraded, due to an intense entanglement with amorphous microfibrils of cellulose (Pauly et al. 1999).

Pectic polysaccharides provide an extraordinary versatile complex, whereby its precise molecular structure has not yet been finally identified. The current status of scientific knowledge was recently reviewed by Ropartz and Ralet (2020). Pectin is composed of as many as eighteen distinct monosaccharides connected through twenty different linkages, while their arrangement is controversially discussed. A broad consensus was found for the smooth and hairy region model postulated by Schols and Voragen (1996), where the unbranched homogalacturonan (HG) domain refers to the smooth region and every branched domain of primary rhamnogalacturonan (RG) and galacturonans forming the hairy region (**Figure 1-1**). The first consists of 1,4-linked α-D-galacturonic acid residues forming a linear unbranched homopolymer chain. In the primary cell wall, HG can be methyl esterified at the C6 position of the galacturonic carboxylic group and acetylated at the C2 or C3 position. Regarding the degree of methylation (DM) and the degree of acetylation (DA), polymers of HG can form gels that increase viscosity, depending on the pH or sugar content of the surrounding media. HG chains occur cross-linked by calcium bridges, where free carboxy groups of HG interact with calcium ions, structurally similar to an "egg-box," determining porosity and fluidity of pectin gel (Grant et al. 1973).

The hairy region mostly consists of RG polysaccharides, bearing a backbone of the repeating disaccharide 1,2-linked α-L-rhamnosyl 1,4-linked α-D-galacturonic acid where acetylation can occur at O-2 and/or O-3. The rhamnose residues have been identified to be substituted with different side chains of neutral polysaccharides at O-4 such as arabinan, arabinogalacturonan (type I and II), and xylogalacturonan. Depending on the nature of the plant, the length of side chains varies enormously ranging from one galactose residue up to chains of 50 residues composed of arabinose/galactose. Side chains also strongly determine the pore size and flexibility of the pectin structure (Ochoa-Villarreal et al. 2012; Schols 1995).

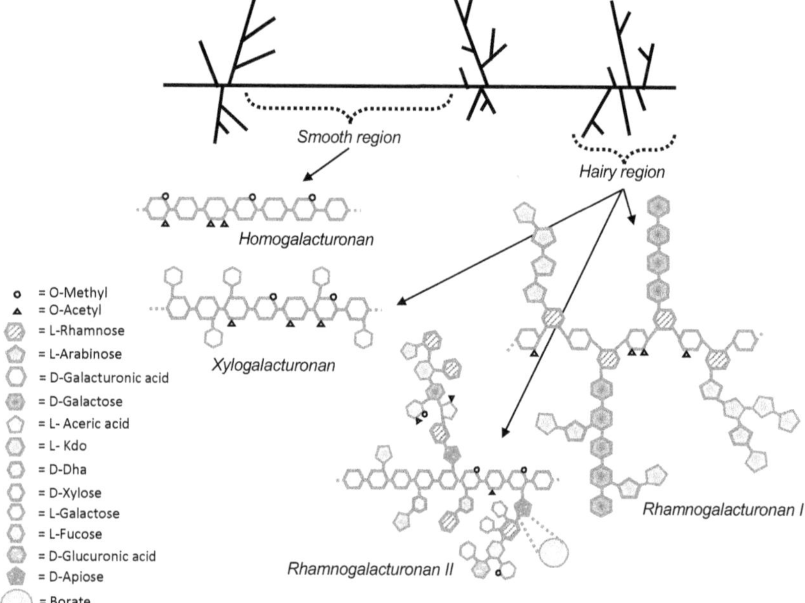

Figure 1-1: Pectin smooth and hairy region model. The smooth region consists of homogalacturonan, and the hairy region includes three types of subunits: xylogalacturonan, arabinan- and type II arabinogalactan substituted rhamnogalacturonan I (RG I). Rhamnogalacturonan II (RG II) is highly branched by five side chains varying in complexity from a single arabinose unit to highly heterogeneous non-saccharides and cross-linked by borate diesters. Ketodeoxyoctonic acid/L-3-deoxy-D-manno-2-octulosonic acid (L-Kdo), L-3-desoxy-D-lyxo-2-heptulosaric acid (L-Dha). Adapted from Harholt et al. 2010 and Ropartz and Ralet 2020.

Another important domain of the hairy region is the highly branched galacturonan (HBG), also termed as RG II. It is a low molecular weight complex (4.8 kDa) with a backbone of nine 1,4-linked α-D-galacturonic acid residues and four different complex side chains. Besides rhamnose, side chains include particularly rare sugars such as 2-*O*-methyl fucose, 2-*O*-methyl xylose, apiose, 3-deoxy-D-manno-2-octulosonic acid (L-KDO), and 3-deoxy-D-lyxo-2-heptulosaric acid (L-DHA), which are especially suitable for the identification. Although the structure of RG II is considered highly conserved throughout different plants, some heterogeneity may occur within these side chains. The RG II subunit is involved in cross-linking of two pectin chains through the borate diester, influencing pore size and stability of the cell wall. Uronic acid residues can chelate cations or occur as methyl esters, thus modulating chelation property (O'Neill et al. 1996; Ropartz and Ralet 2020; Whitcombe et al. 1995).

The basic cell wall structure is similar in all fruit species, but the relative composition of the polysaccharides varies according to the nature of the fruit, tissue origin, environmental factors, status of growing, and the duration of storage influencing fruit firmness and texture (Hilz et al. 2006). Structural changes arise by fruit-born enzymatic activities initiating ripping and degradation processes. Enzymes depolymerize and de-esterify pectic polysaccharides, thereby increasing their solubility, cell wall porosity, and fruit softening during ripening (Jones-Moore et al. 2021). Also, a decrease in calcium concentration in fruit pulp was determined, resulting in the dispersion of conjunction zones by reducing the intercellular network (Stow 1993). However, Ortega-Regules et al. (2008) found that during grape maturation, only small changes in the cell wall with respect to the cellulose, lignin, protein, and total polyphenols content occur, while the concentration of pectin was almost unchanged. The authors explained the latter phenomenon by a simultaneous reduction due to solubilization and active synthesis of, especially, HG.

Due to the complexity of the cell wall built by a very heterogenous pool of polysaccharides, its analytical characterization is a challenging task. Especially the polydispersity and polymolecularity of pectin pose an analytical challenge for detailed characterization. The sequential extraction of the alcohol-insoluble residue of individual polysaccharide fractions by their different solubilities has been established to facilitate the complex cell wall compounds into separated fractions that can be further analyzed.

Until now, only a few researchers focused on the structural characterization and determination of polysaccharides in individual fruits, especially in red berries (Hilz et al. 2005; 2006). Structural characterization of the cell wall polysaccharides of black currant and bilberries revealed a higher pectin content in the former (in detail, HG estimated by the amount of galacturonic acid of the total polysaccharides >45% vs. 24% for black currant and bilberries, respectively). On the other hand, bilberries were characterized by a higher hemicellulose content (revealed by the high amount of glucose and xylose), probably originating from the seeds. Finally, these authors proposed high amounts of HG in berries compared to other fruits; the ratio of HG to RG I, including neutral side chain, is presumed to be approximately 2:1 for all types of berries.

In contrast, the cell wall polysaccharides of red grapes were characterized by a relatively high HG content of pectic polysaccharides (about 80%), while RG II accounted for 1-2% of the berry cell wall. Additionally, pectin subunits of RG I and RG II were enriched three-fold in skin tissue compared to the pulp tissue (Vidal et al. 2001).

Research largely lacks detailed characterization of many berry cell wall profiles. In fact, this basic knowledge provides essential information for choosing the appropriate food production technique considering extraction, polysaccharide degradation, and further polyphenol complexations. Detailed knowledge about the unique cell wall nature of different berries or fruit types allows individually adjusted applications to reveal an efficient cell wall breakage essential for high juice yields and the extraction of sensitive compounds like polyphenols and vitamins, achieving an efficient juice production.

3 Red berry juices

Due to their delicate appearance and fast microbial decomposition, berries are often processed into juices, smoothies, or purees, providing a longer shelf life. Juice production has a long tradition in Germany and has grown to be an important industrial food sector. Germany is one of the leading producers of high-quality fruit juices worldwide with an attained export business of 1.19 billion Euro in 2019 (VdF 2019). Red juices are a highly-priced niche product since the expensive raw materials are susceptible, and production bears different challenges compared to, e.g., apple or orange juice. While red juices of black currant and grape juice are very popular, other intensely colored berries like chokeberries or elderberry have to be blended due to their strong acidity or astringency.

Consumer demand for red juices has grown significantly in the last years since fresh juices provide, similar to the consumption of fresh berries, beneficial health effects. The German Nutrition Society (DGE) and the German Cancer Society (DKG) have strongly promoted fresh juices with various campaigns (DGE 2007). Thus, as emphasized at the beginning, apart from their sensory properties, the quality of red juice is concomitantly defined by its nutritional value. They may also be characterized as a functional food depending on the content of fruit-equivalent amounts of secondary plant metabolites like polyphenols and especially anthocyanins.

The influence of juice processing on the concentration and bioaccessibility of phenolic compounds is discussed controversially. On the one hand, processing ruptures the microstructural elements of the intact berries and potentially releases phenolic compounds into the juice together with other nutrients, leaving them more available for absorption (Jakobek et al. 2007). On the other hand, phenolic compounds, such as anthocyanins, are easily oxidized and susceptible to degradative reactions during processing and storage. Thus, the final juice has often been shown to contain fewer valuable compounds than the intact fruit. For instance, in blueberry juice, only 36-39% of polyphenolics and 13-32% anthocyanins from the original amount in berries were recovered in the pasteurized juices (Lee et al. 2002; Skrede et al. 2000). The anthocyanin retention in cranberry juice was described as even generally <50% (Pappas and Schaich 2009). However, it has to be considered that polymerization effects might decrease anthocyanin concentration, resulting in reputedly lower anthocyanin values, due to the hampered analytical detection.

Although processed berry juices contain a satisfactory amount of polyphenols, substantial quantities were found to be separated with skin cells and press residues (Hilz et al. 2005; Landbo and Meyer 2001; Wrolstad et al. 2005). Press residues contain mainly plant cell polysaccharides encompassing cellulose, hemicellulose, pectin, and lignin, but also proteins, stabilized by strong ionic and covalent linkages. This highly complex framework presents a barrier to the release of polyphenols and hinders their extraction during processing. Additionally, the degraded cell wall residues of high molecular polymers might complex with released polyphenols to insoluble complexes, also remaining in the press cake (Leong and Oey 2017).

In general, the more processing steps were applied for the production of red juice, the greater was the loss of total polyphenolics, total anthocyanins, and vitamin C compared to their clarified, filtered, or pressed juice controls in several studies (Leong and Oey, 2017). As a consequence of all processing steps, the genuine concentration and profile of anthocyanins are significantly altered, with significant impacts on juice color and nutritional value (Brownmiller et al. 2008; Holzwarth et al. 2012; Maier et al. 2009). While the reduction of the red-violet color adversely affects consumer acceptance, lower polyphenol concentrations decrease potential health benefits compared to the native fruits.

After processing, polyphenol degradation, including anthocyanin degradation, continues during storage. Here, light and temperature play an important role. For example, blueberry juice stored at 23 °C for 60 days only retained 50% of anthocyanins (Srivastava et al. 2007). This remains the main reason why it is suggested that fruit juices must be consumed within 48 hours after opening by most companies (Patras et al. 2010).

The production of fruit-like red juices with comparable compositions to the original fruit and long shelf lives needs to be improved. A lot of research has already been conducted to clarify and reduce the deteriorative impacts of individual juice production steps on the polyphenol and anthocyanin concentration, the latter of which has been reviewed in detail by Weber and Larsen (2017). However, processing improvements by *inter alia* the implementation of novel technological approaches require additional basic research to obtain comprehensive knowledge about quality-determining mechanisms on a molecular level during red juice production. Understanding cell wall degradation, polyphenol binding, and complexation matters are important for extraction purposes and essential for shelf life.

4 Production of red berry juice

Due to the specific nature of different fruits, juice production has already been adjusted individually to overcome their distinct challenges. Accordingly, appropriate steps have been implemented in the process of the conventional red berry juice production. **Figure 1-2** displays a general schematic overview of the berry juice production process.

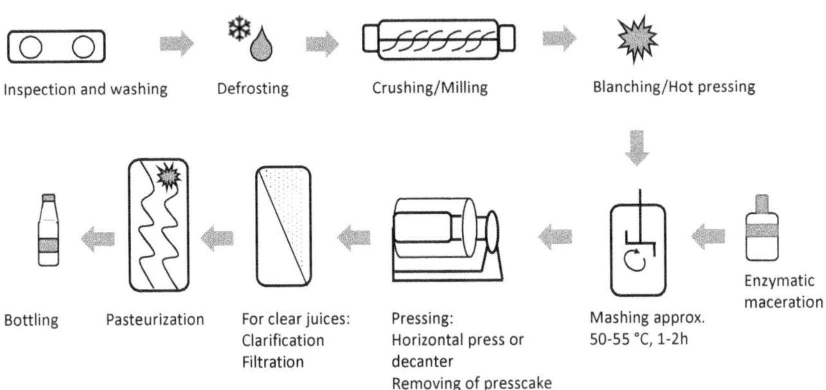

Figure 1-2: Schematic overview of the conventional red berry juice production modified after Grassin and Coutel (2010).

Berries are very sensitive to mold contamination, especially by *Botrytis cinerea*, which leads to fast juice browning by polyphenol oxidase production, hinders juice extraction, and fouling of filters. An initial heating of the raw material (>80 °C), also termed as blanching, inactivates exogenous and fruit-own oxidases, reducing oxidative discoloration and browning issues preemptively. Concomitantly, the short and intense heat treatment is discussed to facilitate pigment extraction due to thermally induced permeabilization of the cell wall structure, which results in more color-intensive juices (Leong and Oey 2017).

However, heating treatments usually decrease the concentration of anthocyanins by thermal hydrolysis. Correspondingly, the effects of pasteurization (commonly at 90 °C) demonstrate an overall decrease in anthocyanin concentration depending on processing parameters, food matrix, and anthocyanin structure. Optimized pasteurization parameters conducted in several studies show that the shortest treatments (e.g., 30 s at 90 °C for strawberry juice pasteurization by Odriozola-Serrano et al. 2008) revealed the lowest loss in anthocyanin concentration. Matrix compounds like dissolved protein residues, colorless flavonoids, and polysaccharide residues have been found to mitigate the destructive processing effects to anthocyanins due to the formation of stabilizing complexes like copigmentation (Padayachee et al. 2012; Sadilova et al. 2009).

The most crucial challenge in red juice production is probably the one caused by the firm skin of the berries as a result of the pectin-enriched cell wall matrix. The ratio of skin to berry flesh tissue is comparatively high because of their small size. Heating dissolves polysaccharides from pectin into the mash, causing a higher viscosity and thus impeding juice extraction and filtration. This issue has been solved by the addition of pectinolytic enzymes during mashing, referred to as enzymatic maceration, which will be introduced in more detail below. The enzymatic cleavage ruptures the pectinolytic cell wall, reduces viscosity, and allows for sufficient juice yields.

Recently, alternative cell wall-degrading techniques have successfully been introduced into the juice industry with growing usage worldwide. The primary objective of novel technologies is to generate high-quality juices by maintaining their color, aroma, and nutritional value via gentle process conditions by primarily minimizing heat-intensive processing. These techniques are also termed as “green technologies” and declared as environmentally friendly and sustainable due to their reduced time, water, and energy requirements, thereby also increasing economic and ecologic advantages. Of particular note in juice production are the novel technologies of ultrasound, pulsed electric fields (PEF), high-pressure processing (HPP), cold plasma processing (CP), and flash vacuum extension (Knorr, 2011; Weber and Larsen 2017). The latter authors also took a closer look at the influence of the respective technologies on anthocyanin profile and concentrations in berry products. These techniques mainly rupture plant cell walls mechanically and accelerate mass transfer from the inner cell matrix, thus improving juice yield and extraction of polysaccharides and polyphenols by concomitantly less heat-induced off-flavors or unattractive color changes. The main advantages of these techniques are that they serve as non-thermal pasteurization alternatives, while their application ensures sufficient inactivation of microorganisms and enzymes. With regard to polysaccharide degradation, US was reported to have the most potential by producing polymers of the lowest molecular weights, while the effects of HPP, PEF and, CP were less pronounced (Cui and Zhu 2021). In this thesis, a focus was taken on implementing ultrasound technology into red juice production as an alternative method for cell wall disintegration (**Section 4**). A great advantage compared to the other techniques is the simple setup and low-cost equipment, which can easily be installed in research lab conditions and scaled up to industrial production lines.

While much research has been conducted on an optimized juice or polyphenol yield, process-induced effects on other relevant compounds, such as polysaccharides, have often not been considered. Indeed, cell wall polysaccharides are not only target substrates for the applied enzymes but also interact with polyphenols, changing their profiles and contents (Buchweitz et al. 2013a; Jakobek 2015; Renard et al. 2001). The implementation of novel cell wall degrading techniques further increases the diversity of polysaccharide degradation and their possible effects on other matrix compounds such as anthocyanins.

Because juice quality is strongly influenced by the resulting pool of polysaccharides and oligosaccharides, this thesis focuses on the characterization of the polysaccharide degradation during mashing. Thereby, the degradation of pectin by the commonly enzymatic maceration is compared to the novel ultrasound-assisted maceration (**Chapter 2**). Pectin degradation has further been predominantly considered regarding possible anthocyanin complexation properties, influencing anthocyanin concentration, profile, and stability (**Chapter 3**).

4.1 Enzymatic maceration

Enzymatic maceration significantly determines juice yield and the extraction rate of polyphenol compounds during juice production. Since 1930, industrial enzymatic preparations obtained from microorganisms have successfully been applied in juice processing. During mashing, cell walls are disrupted mechanically, and fruit-borne enzymes are released, degrading protopectin mainly from the middle lamella into soluble pectin fragments, forming a highly viscous gel that is hard to press. The addition of exogenous enzyme preparations accelerates cell wall degradation, resulting in decreased viscosity, improved juice yield, and reduced production time (todays standard maceration time as of 1-2 h) (Grassin and Coutel 2010). For instance, the yield of black currant juice was improved from 46-65% to over 80% by enzyme addition during maceration (Laaksonen et al. 2012).

In the European Union (2001/112/EC, 2002), treatments with pectinolytic, proteolytic, and amylolytic enzymes are permitted for fruit juice production. Currently, enzyme suppliers offer specific enzyme preparations tailored to the special conditions of juices prepared from each kind of berry. These commercial enzymes are able to act at high temperatures and high fruit acidity, which otherwise are common inhibiting parameters that enzymes are affected by. The preparations consist of fungal pectinase enzymes containing multiple pectinolytic activities produced predominantly by *Aspergillus niger*.

Besides the three main pectinases, which are polygalacturonase (PG, exo/endo), pectin methylesterase (PME), and pectin lyase (PL), commercial enzyme preparations applied for the production of red juices also contain minor activities from enzymes such as rhamnogalacturonase and rhamnogalacturonan acetyl esterase. Due to their conventional biotechnological production, numerous side activities like glycosidases, galactanases, arabinases, hemicellulases, cellulases, or proteases are additionally included that degrade pectin and its associated molecules sufficiently (Grassin and Coutel 2010; Whitaker et al. 2003). Enzymatic activities are defined and classified based on their actions on pectin and grouped into three main categories of lyases, hydrolases, and esterases. While lyases and hydrolases depolymerize the pectin main chain of HG, esterases are active on methyl esters and acetyl esters of galacturonic acid moieties of HG and RG structures (Grassin and Coutel 2010; Whitaker et al. 2003). Consequently, the complete degradation of pectin requires an ensemble of all the above-mentioned pectic enzymes (de Vries and Visser, 2001).

Although enzymatic maceration has a long tradition in juice production, currently, commercial enzyme preparations have only been designed mainly to optimize juice yield and avoid haze formations. As mentioned above, information on detailed polysaccharide profiles of most berry cell walls is missing so far. Nevertheless, structural information on the cell wall polysaccharides of cellulose, hemicelluloses, and pectins could provide an optimal enzymatic system based on just this information. A corresponding first approach was applied by Hilz et al. (2005), who characterized the polysaccharide profile of intact bilberries and blackcurrants, continued at juice production via enzymatic maceration by taking into account the composition of the cell wall polysaccharides. Results demonstrated that not only enzymatic degradation but also the natural composition of polysaccharides of the distinct fruits have a tremendous impact on the oligosaccharide composition of the final juice. Additionally, Ducasse et al. (2011) reported that similar enzymes showed varied polysaccharide hydrolytic patterns on pectin obtained from different grape varieties thus releasing a diverse pool of polymers. These findings again imply the need for a detailed characterization of individual berry cell walls to extend the knowledge on efficient cell wall degradation that can be controlled by, *e.g.,* the selection and dosage of appropriate enzymes. Thus, tailormade enzyme preparations could be designed explicitly by a more profound knowledge of the fruit polysaccharide structure. Additionally, this knowledge may also contribute to the regulation and prediction of polysaccharides arising during production, which is important to control juice quality.

4.2 Polysaccharide modifications

During an optimal enzymatic maceration, enzymes must disintegrate most of the pectic polysaccharides but not dissolve the whole cell wall matrix into neutral colloids. The latter would adversely affect clarification and stabilization. Undissolved polysaccharides are separated after the maceration process by pressing. The resulting pomace includes cell wall residues like pectin, hemicellulose, cellulose, lignin, and proteins.

The unfiltered juice obtained contains pectic polysaccharides with a low molecular weight (up to 1 Mio. Da). Interaction of polysaccharides with proteins and polyphenols results in the formation of cloudiness and haze. To eliminate this, a second enzymatic treatment can be applied to obtain clear juices, followed by separation through filtration and fining. Nevertheless, small quantities of soluble oligosaccharides (MW <10,000 Da) still remain in the clarified juice (Kressmann 2001). For instance, soluble colloids in grape must consisted of 60% soluble sugars; the rest was attributed to proteins, polyphenols, and salts (Guadalupe and Ayestarán 2007). In juices, final polysaccharide concentrations of up to 7,000 mg/L in black currant juice and up to 300 mg/L in grape juice have been reported depending on different production procedures (Ducasse et al. 2011; Kressmann 2001; Mieszczakowska-Frac et al. 2012; Renard et al. 2001).

Certain polysaccharides are also considered as protective colloids, as they help in the prevention or limit aggregation and flocculation. Additionally, they contribute to the organoleptic properties of juices and wine, stabilizing flavor, color, and foam (Ayestarán et al. 2004; Guadalupe and Ayestarán 2007; Vidal et al. 2004). Although these polymers strongly contribute to product qualities, the exact molecules responsible for this and the binding mechanisms are still unclear and remain as goals of many recent investigations.

Soluble polysaccharides have been identified as HBR/RG II dimers, while RG I residues have been defined as modified hairy regions (MHR) that are resistant to degradation by pectinolytic enzymes applied. RG II dimers generally occur as borate complexes and have been found to be the dominant polysaccharides in several juices after enzymatic liquefication (Doco et al. 1997). One-third of total juice polysaccharides were referred to RG II residues in the case of bilberry and blackcurrant juice (Hilz et al. 2005). In red wine, >50 mg/L of RG II residues (20-46% of total soluble polysaccharides) were quantified, demonstrating even longer stability over a period of ten years (Ayestarán et al. 2004; Doco and Brillouet 1993; Pellerin et al. 1996). MHRs consist of branched RG I units attached to neutral side chains of arabinan, arabinogalactan, and galactose, and also of minor HG residues (Ducasse et al. 2011; Hilz et al. 2005). Additionally, high contents of arabinogalactan type II, arabinogalactan proteins (AGP), and HG residues consisting of 10-15 galacturonic acid residues were identified in clarified juices and wines (Ayestarán et al. 2004; Ducasse et al. 2011; Kressmann 2001).

The fact that the structure of these soluble polysaccharides differs widely, depending on the fruit type, ripening status, and the enzymes applied, various complications in their characterization and the evaluation of an underlying degradation pattern are often encountered. Simultaneously, prediction of the concentration, structure, and composition of the resulting oligomers and polymers also becomes very challenging. The present dissertation focuses on the characterization of pectic polysaccharides and oligosaccharides after enzymatic degradation contributing to the knowledge of pectic profiles similar to a berry juice maceration (**Chapter 2 and 3**).

4.3 Extraction and complexation of polyphenols including anthocyanins

Cell wall disintegration during enzymatic maceration not only improves juice yield but also increases the amount of polyphenols such as flavonoids released into the juice. Especially water-soluble anthocyanins belonging to cell wall-bound polyphenols and non-cell wall polyphenols (confined in the vacuoles of plant cells or associated with the cell nucleus) become more extractable after cell wall degradation (Acosta-Estrada et al. 2014; Pinelo et al. 2006). After a treatment with enzymes, the content of total anthocyanins in blackcurrant juice was increased by 68% (Landbo and Meyer 2004) and 41% for bilberry juice (Buchert et al. 2005), in pomegranate juice, the total phenolic content was raised by 125% (Rinaldi et al. 2013).

Although enzymatic maceration enhances polyphenol and anthocyanin extractability, a considerable amount still remains in the press cake, as observed in several studies, e.g., in blueberry press cake 15%-55% (Brownmiller et al. 2008; Lee et al. 2002; Skrede et al. 2000), in blackberry press cake 14% (Hager et al. 2008), and in chokeberry press cake 22% (Wilkes et al. 2014). The extractability of anthocyanins depends on multiple factors, like the fruit ripening state (Jones-Moore et al. 2021), the enzymatic degradation of the cell wall matrix (Bagger-Jørgensen and Meyer 2004), the polysaccharide amount, and its composition within the cell wall (Ortega-Regules et al. 2006), and the individual structure of the anthocyanins (Koponen et al. 2008). Consequently, berries rich in pectin, such as black currants, require a higher dosage of pectinolytic enzymes to obtain similar anthocyanin yields as observed for berries that are not so concentrated with pectin, like bilberries (Koponen et al. 2008). Indeed, enzyme dosage is limited since overdosage may facilitate undesired enzymatic side effects like β-galactosidase activity that is assumed to remove sugar residues from anthocyanins, strongly decreasing their stability (Buchert et al. 2005; Heffels et al. 2017; Landbo and Meyer 2004). For this purpose, the demand for alternative degradation techniques enabling higher anthocyanin extractions is of great interest and is thus introduced in **Section 4**.

The mechanisms of polysaccharide-anthocyanin complexation are not only important for extractability purposes, where interactions have to be severed so that anthocyanins are released. They also account for the very likely formation of new complexes between the extracted anthocyanins and polysaccharide fragments during juice production resulting from the cell wall disintegration.

Considerable binding of polyphenols to different cell wall components is well documented with the help of various experimental models. Synthesized substructures in these models show saturated binding concentrations of 0.3-1.5 g adsorbed polyphenols/g dry weight of polysaccharide material (Phan et al. 2017). The adsorption occurs very selectively since it is influenced by several aspects like structure and composition of both counterparts, which are introduced subsequently.

In red juices and red wine, soluble polysaccharides like MHR and RG II have been described as intensely red-colored due to anthocyanin complexation. In black currant and bilberry juices, these polysaccharides were associated with almost 30% of the total polyphenols (Hilz et al. 2005; 2006). The complex formation is suspected to protect anthocyanins against oxidation or other degradative reactions, thus prolonging their stability. For instance, clarified strawberry and black carrot juices showed a significantly lower anthocyanin stability compared to the untreated controls, which may be explained by the pectin removal (Sadilova et al. 2009). Subsequently, pectin-enriched juices increased anthocyanin half-life values. In the case of berry jams, pectin-based gels revealed remarkably high anthocyanin stabilization compared to other typical gels used in jam production (Holzwarth et al. 2013; Kopjar et al. 2009; Poiana et al. 2013).

Until today, only little research has focused on the interaction mechanisms of pectin and anthocyanins in particular. However, some research has been conducted to evaluate interactions between cell wall polysaccharides and various polyphenols using model systems (Liu et al. 2020). The findings describe the driving force for polyphenol-polysaccharide interactions by reversible non-covalent interactions, mainly hydrogen bonds, hydrophobic interactions, and van der Waals forces.

Pectin has been shown to have a remarkably high affinity for polyphenols than most other polysaccharides in solution or suspension, especially for catechin (Liu et al. 2019), proanthocyanins (Le Bourvellec et al. 2005), and anthocyanins (Padayachee et al. 2012). Additionally, complexations are influenced by numerous factors pertaining to both polyphenols and polysaccharides. The former have been demonstrated to affect binding as a result of their molecular weight, galloylation and hydroxylation, methylation and acetylation, glycosylation, and the molecule's stereochemistry with respect to the stereochemical pyran rings of flavan-3-ols, as seen in catechins or epicatechins. The latter were shown to affect the interaction by their botanical origin (cellulose, hemicellulose, pectin), the surface area, porosity, the degree of esterification of pectins, branching complexity, molecular weight, solubility, and the presence of other molecules in the cell wall (Liu et al. 2020).

The interaction between anthocyanins and pectin, in particular, is assumed to be mainly driven by hydrogen bonds built by hydroxy groups of the anthocyanin B-ring and free galacturonic acid residues from the HG chain of pectin **(Figure 1-3)** (Buchweitz et al. 2013a; 2013b; Fernandes et al. 2015; Holzwarth et al. 2012). It has been demonstrated that a higher number of hydroxy groups on the B-ring can promote adsorption (Fernandes et al. 2015). On the other hand, a higher number of free galacturonic acid residues in the pectin molecule increased anthocyanin complexation by providing more binding sites (Kopjar et al. 2009).

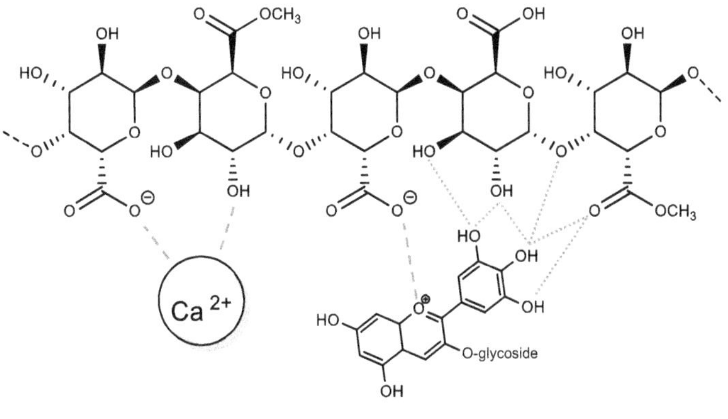

Figure 1-3: Schematic illustration of the interaction between pectic galacturonic acid residues and the anthocyanin cation (delphinidin) similar to the calcium ion association in acidic medium. Ionic interactions (dashed lines) occur between charged ions, hydrogen bonds (dotted lines) occur between the hydroxy groups of both counterparts (modified after Buchweitz et al. 2013).

Besides the direct interaction between anthocyanins and pectin, a second, slower adsorption mechanism by self-copigmentation has been observed. Here, anthocyanins associate via intermolecular π-π-stacking on the initially absorbed anthocyanin molecule by hydrophobic interactions and dispersion forces during longer exposure (Medina-Plaza et al. 2020; Padayachee et al. 2012; Renard et al. 2001).

In slightly acidic media (pH 3-4), as observed in berries and their derived juices, ionic interactions additionally facilitate the interactions between anthocyanins and pectin. A unique feature of anthocyanins within the polyphenols is that they occur as positively charged flavylium cations in acidic environments due to their pH-dependent structural equilibrium. Concurrently, the polygalacturonic acid (pK_a 3.5) of pectin is negatively charged in the slightly acidic milieu of juices. A very clear illustration of the resulting interaction mechanisms between anthocyanins and the HG domain of pectin was postulated by Buchweitz et al. (2013), and has been adapted in **Figure 1-3.** Similar to the egg-box model, electrostatic interaction is drawn between the negatively charged dissociated carboxylic groups and the anthocyanin flavylium cation as the counterpart to the calcium ions.

Although the egg-box model has been hypothesized as the interaction mechanism quite often, there are some pieces of evidence for additional interaction forces that have not been entirely understood so far. The same authors could not explain different anthocyanin stabilizing effects of various pectin types from apple, citrus, and sugar beet based on their binding theory (Buchweitz et al. 2013b). While the egg-box model only describes the interaction between anthocyanins and the HG domain of pectin, other observations indicated interactions toward the neutral side chains of the pectin hairy regions (Mazzaracchio et al. 2004). These divergent findings might be caused by the usage of commercially extracted pectin containing $\geq$65% galacturonic acids indicating, high amounts of HG in many studies. In contrast, native pectin, in particular of berries, includes a certain amount of neutral monosaccharides related to hairy regions. These subunits form a voluminous porous complex, providing hydrophilic domains and hydrophobic pockets, where polyphenols can be encapsulated. In the case of non-polar polyphenols, hydrophobic interactions are assumed between their aromatic π-electron system towards the non-polar sites of polysaccharides (Liu et al. 2020; Ropartz and Ralet 2020).

Despite the importance of polyphenol yield and their stability as the key quality aspects of organoleptic and nutritional properties in juices, the mechanisms that finally drive these interactions are still not well understood. Since berry cell walls appear in a more heterogeneous complex structure than commercially extracted pectin, reactions during juice productions are more multifaceted and complicated. Because numerous factors affect the interactions, the understanding of detailed mechanisms is demanding and requires more research. For this purpose, this dissertation investigates the interaction of several anthocyanins varying in either their sugar moiety or aglycone with also structurally diverse pectic polysaccharides to gain more knowledge on structural dependencies for binding mechanisms (**Chapter 3**).

Apart from this, process-induced effects like the polysaccharide modifications by enzymes have rarely been considered. Enzymatic treatments of pectin-rich strawberry purées by pectin methylesterase (PME) and polygalacturonase (PG) showed higher anthocyanin retention after storage for eleven weeks (Holzwarth et al. 2012). Authors assume that enzyme-treated pectin led to much higher retention of anthocyanins due to liberated galacturonic acid binding sites. The authors point out that PG treated purée may also contribute to the co-pigmentation effect resulting from the enhanced release of pigments and co-pigments due to cell wall degradation (Holzwarth et al. 2012). Apart from adverse β-glucosidase side activities acting directly on anthocyanins, only little is known about the impact of enzymatic maceration that can indirectly stabilize anthocyanin concentration by appropriate pectin modification. Therefore, this dissertation considers the binding ability of enzymatically modified pectin with several anthocyanins to investigate complexation similar to the mashing step during juice production (**Chapter 3**).

In conclusion, the effects of enzymatic maceration on juice quality are tremendous. Cell wall degradation still has tremendous potential to be improved considering polyphenol extraction is far below the maximum yield, as proven by the high amounts of polyphenols that remain in the press cake. It is essential to consider the resulting oligosaccharides and polysaccharides as they greatly determine the juice quality by influencing organoleptic properties like the formation of haze and polyphenol complexation, thereby affecting color and flavor. To overcome these demanding challenges, the application of ultrasound technology has been proposed with promising effects, including several aspects introduced in the following section.

5 Ultrasound

The potential of ultrasound in food processing was first presented in the 1920s with the purpose of cleaning and homogenization. To date, ultrasound application is highly versatile, and ultrasound-assisted processes have been described for nearly every aspect of food production, including crystallization, degassing, drying, extraction, filtration, freezing, homogenization, emulsification, meat tenderization, and microbial inactivation (Knorr 2011). Besides the high efficiency, the advantages of ultrasound also include sustainable energy saving potential due to enhanced mass and heat transfers, decreased processing time, reduced temperatures, selective extractions, and increased productivity, accompanied by comparably low instrument requirements.

5.1 Ultrasound setup and mechanisms

Ultrasound waves are generated by an ultrasound transducer which transfers electric energy into mechanical vibration by piezoelectricity. The so-called sonotrode or probe directly transmits the vibration into a liquid medium propagating longitudinal waves with continuous compression and rarefaction. At this, the liquid quantity in the overall sample should be at least 5% (Feng et al. 2011). The power input of the transducer determines the frequency and is directly proportional to the maximum amplitude of the sinusoidal waves. The ultrasonic energy transmitted by the medium can be defined by the ultrasound power (W), ultrasound intensity (W/cm^2), or ultrasound density (W/mL). The latter is strongly influenced by the volume of the whole liquid medium and therefore has been broadly accepted as the most proper dimension. In the context of juice production, high intensities referred to as power ultrasound are required that are generated by large amplitudes to operate with acoustic frequencies ranging from 20-100 kHz (Feng et al. 2011).

There are two general designs of ultrasonic devices that are employed in the industry: A single transducer placed in a titanium cylinder called an ultrasonic probe or horn and an ultrasonic bath. The probe produces an intense acoustic field directly below its surface (amplitude of approximately 100 µm), where there is a possibility to control the energy input and intensity very precisely. The corresponding input is defined by the treated volume and time exposure (kWh/L·s), while the power output is given as a function of the surface area of the sonotrode (W/cm^2). Conversely, the ultrasonic bath distributes a more diffused energy field, providing a much lower energy intensity. Responsible for that are several distributed transducers across the base of the bath. In the context of an industrial juice production line, the ultrasonic probe is more practical due to the possible application into a flow cell of a continuous system. Since ultrasound strongly increases temperature, a continuous system allows for better control of temperature because of the short contact of the treated sample. In **Chapter 2,** a continuous flow application was designed to meet industrial requirements for a practical ultrasound-assisted maceration approach. Since energy and intensity are scale-independent, also minor ultrasonic approaches may reveal valuable knowledge about process conditions and their effects (Kentish and Feng 2014).

Ultrasound effects strongly depend on the energy applied. The transmission of ultrasound waves causes high vibrations and pressure fluctuations into the medium, which leads to an enhanced mass transfer, homogenization, dispersion, disintegration, and cavitational phenomena. The latter are responsible for most ultrasonic-induced effects of high-power ultrasound in food processing. Cavitation not only facilitates mechanical effects but also induces chemical and biochemical reactions. Cavitational microbubbles are formed during the rarefaction cycle of the sound waves by air inclusion due to the incompressible liquid medium that cannot readily accommodate the rapid pressure changes. The frequencies of the transducer determine the bubble size. Cavitational zones may contain thousands of many such bubbles (Kentish and Feng 2014). Once formed, bubbles grow by rectified diffusion or undergo coalescence. The motion of stable bubbles leads to microstreaming and enhances fluid mixing. When reaching a specific size range, the negative pressure inside the bubbles becomes higher than the medium's surface tension; thus, the bubbles collapse violently in a microscale implosion, rising local temperatures to 5000 K and pressures to 1,200 bar. Bubbles can implode destructively into daughter droplets or undergo a sequence of implosions. Accordingly, these cavitational collapses generate shockwaves and turbulences into the medium intensifying mechanical effects. The local temperature rises accompanied by strong shear forces that are termed as hot spots and generate fluid high-speed microjets providing immense destructive energy near a cell wall surface. This effect explains the ultrasound-assisted enhanced extraction rates of plant compounds like polysaccharides or polyphenols, where massive tissue destruction increases the permeability (Kentish and Feng 2014).

Ultrasound frequency is inversely proportional to the bubble size. Low frequency generates large cavitation bubbles resulting in higher temperatures and pressures in the cavitation zone, accounting for power ultrasound. As the frequency increases, more collapse events occur per time, resulting in a uniform, albeit less intense, sound field until the cavitation zone becomes less violent due to non-cavitation collapses. In this case, mixing and shear forces predominate generated by the acoustic streaming. Generally, an ultrasound wave with 20-1000 kHz frequency possesses enough energy to break down fruit matrices and facilitates extraction. The ultrasound amplitude level is also an indicator of the cavitation strength and is often used to represent the ultrasound power (Kentish and Feng 2014).

Besides the mechanical effects, cavitation causes chemical and biochemical effects. The first is triggered by the ultrasound ability of water dissociation into dissolved oxygen molecules, creating highly reactive free radicals (•H, •OH, •OOH). Reactive species can further form hydrogen peroxide, recombine or react with other molecules. The generation of free radicals is positively related to ultrasound intensity and treatment duration but negatively correlated to temperature. Although the chemical effects are more pronounced at higher-frequency ultrasound in a field referred to as sonochemistry (>100 kHz), it also has been observed less pronounced under lower intensity treatments (Ma et al. 2020).

Biochemical reactions are related to structural changes of bioactive compounds such as proteins acting as enzymes and polysaccharides serving substrates. Since many food matrices are rich in both classes of these molecules, the related changes are highly relevant in food production and will thus be discussed below.

Ultrasound effects have been demonstrated to be highly variable depending on ultrasound equipment and design, encompassing intensity, energy input, temperature, pressure, and various properties of the food matrix, including density, viscosity, and vapor pressure of the liquid. Therefore, they have to be considered carefully for every particular application.

5.2 Ultrasound in fruit juice production

Two ultrasound-assisted approaches are part of current research – an alternative non-thermal pasteurization technique and a milder maceration method encompassing an enhanced extraction rate. Both purposes aim at producing high-quality juices with respect to high retention of nutritionally valuable compounds like polyphenols and a naturally fresh-like taste and color. Regarding pasteurization, several studies demonstrated a successful reduction in microbial counts or inactivation of microbial and enzymatic activities (Li et al. 2017, Paniwnyk et al. 2017, Zinoviadou et al. 2015). Thus, high-intensity ultrasound treatments have been shown to significantly enhance shelf life. However, ultrasound application requires the combination of pressure or temperature, referred to as manosonication or thermosonication, respectively, to ensure microbial safety of juice products. The advantage of these techniques is mainly caused by shortened treatment time and lower thermal loads with concomitantly less undesired side effects in the context of taste, color, and antioxidant activities compared to conventional pasteurization techniques.

Research on ultrasound-assisted maceration is only in its infancy. Thus, the number of studies is limited so far. Nevertheless, the existing studies present promising results in the cases of grape, mulberry, or guava must, where the combined application revealed the highest phenolic and flavonoid contents, total acidity, and reducing sugar contents compared to the exclusive treatments of both degrading methods in shortening maceration time (12-13 min vs. 60-90 min as conventional maceration time). However, studies strongly differ in their maceration conditions in case of equipment (bath or horn), intensities (60-360 W), optimum temperature (20 °C to 74 °C), and the enzyme preparation applied (pectinases, cellulases) (Dalagnol et al. 2017; Lieu and Le 2010; Nguyen et al. 2013; Tchabo et al. 2015). Since all these parameters strongly influence the output, more research is needed to generate a consistent knowledge of optimized treatment conditions, and to examine the effects and mechanisms of the ultrasound-assisted enzymatic degradation. Studies covering individual effects of this technique on red berry mash compounds have been published, although keeping different research objectives in context. In the following, a brief review of related findings is given about polyphenols (anthocyanins), polysaccharides (pectin), and enzymes (pectinase).

5.2.1 Ultrasound effects on anthocyanins

Surprisingly, the otherwise very vulnerable anthocyanins have been proven to be relatively stable under sonication conditions during extraction. Even though minor losses, probably due to temperature increase and oxidation by ultrasound-induced hydroxyl radicals, are reported after treatments of several red berry juices, the overall retention was mostly above 90%. Increases in anthocyanin concentration were often observed and explained by the effective removal of occluded oxygen from the juice or superior extractions from suspended pulp particles in cases of unclarified juices. Additionally, the inactivation of oxidative enzymes like polyphenol oxidases was discussed for improved anthocyanin retention. Polyphenol extraction was, in general, significantly enhanced by 6-35% compared to conventional extraction methods, due to the disintegration of cell wall increasing permeability and improving mass transfer (Vilkhu et al. 2008).

5.2.2 Ultrasound effects on polysaccharides

Ultrasound has been applied to facilitate the extraction and modification of structural and physicochemical properties of various water-soluble polysaccharides. Since oligo-saccharides have been found to be related to the improvement of metabolic functions, decreased blood lipid levels, control of cardiovascular diseases, and prebiotic effects, the extraction and precise modification of these compounds have been focused on current investigations.

In general, ultrasound has proven to significantly reduce the molecular weight of polysaccharides, depending on their type, original size, and concentration. While lower intensities lead to dispersions of weak interactions like hydrogen bonds or hydrophobic interactions and thus entangled polymer complexes, higher intensities break glycosidic bonds. The resulting degradation products appear in a narrow uniform molecular weight distribution, indicating a non-random breakage mechanism (Yan et al. 2016). The detailed mechanisms of polysaccharide and, in particular, pectin extraction, degradation, and modification by ultrasound have been reviewed by Cui and Zhu (2021), Wang et al. (2018), and are discussed in **Chapters 2 and 4**. Extraction rates and degradation can be controlled by ultrasonic intensity, frequency, treatment time, and temperature. Thus, it is recommended to adjust each ultrasound setup individually, considering purpose and sample properties. For example, gentle conditions allow the extraction of natural pectin with only slight structural alterations.

Although ultrasound has been shown to achieve comparable reductions in molecular weight similar to conventional extractions, the respective ultrasound parameters required considerably harsher conditions (>100 W/cm) compared to the aforementioned polyphenol extraction and treatments (<<100 W/cm) (Carrera et al. 2012; Wang et al. 2018). Plausibly, the distinct conditions have been adjusted individually to the particular purpose of each research goal, disregarding matrix compounds. For the purpose of a comparable enzymatic cell wall degradation during juice maceration, these harsh conditions might be detrimental to polyphenols, while lower ultrasound energies will likely not cause the desired cell wall breakage into soluble polysaccharides. However, lower ultrasound intensities (~10 W/cm) have been demonstrated to enhance polysaccharide degrading enzyme activities by synergistic effects used as ultrasound-assisted enzymatic extraction (Ma et al. 2016; Nadar and Rathod 2017).

5.2.3 Ultrasound effects on enzymes

In the last decade, ultrasound has extensively been used in enzyme-catalyzed reactions and biotransformation in numerous fields, aiming to improve enzymatic efficiency in the reaction processes by reducing operating costs and increasing output. Ultrasound-assisted applications have been reported for industries of textiles, pharma, oleo-chemicals and detergents, food and perfumery, cosmetics, and also in organic enzyme catalysis (Nadar and Rathod 2017).

Despite the many successfully established applications of ultrasound-assisted enzymatic extraction, the underlying mechanisms are not fully understood so far. Nadar and Rathod (2017) and Wang et al. (2018) provided an overview on several factors depending on the degradation reactions, which reveal destructive or supporting synergistic effects. Besides an enhanced mass transfer of the enzyme-substrate diffusion, the conformational modification of substrates and enzymes is discussed that facilitates enzymatic cleavage simultaneously. Ultrasound treatments entangle and degrade polymers to simpler molecules. As a result, these polymers are easier accessible to the enzyme binding sites that are concomitantly more exposed by ultrasound-induced enzyme conformation changes. Additionally, it is assumed that the released cavitation energy warms up the medium, which may prove to be beneficial for enhanced enzyme activity. However, synergistic effects strongly depend on ultrasound intensity and the type of enzyme. Higher ultrasound intensities cause enzyme inactivation by denaturation desired in non-thermal pasteurization. Thus, the experimental setup and ultrasound parameters have to be carefully adjusted for each purpose. Research devoted to the ultrasound-assisted enzymatic modification of cell wall polysaccharides, in particular pectin, is still rare and varies enormously in the experimental conditions applied, complicating comparability.

Combining all these constructive features of ultrasound-assisted research that was introduced in the last paragraphs, ultrasound technology provides a promising technique to establish gentle cell wall breakdown and polysaccharide degradation during enzymatic maceration, considering protective complexation properties and polyphenol susceptibility. However, ultrasound effects strongly depend on many aspects that have to be considered all together for designing an industrially practicable maceration process. Thus, **Chapter 2** encompasses a fundamental contribution to the investigation of ultrasound-assisted enzymatic maceration by considering the effects on pectin degradation that serves to be highly important during fruit juice maceration.

6 Aims of the thesis

The quality of red juices is strongly determined by their amount of anthocyanins since they are responsible for the juice's nutritional value and organoleptic properties. The final anthocyanin concentration greatly depends on juice production, and more specifically, on the cell wall breakage during maceration. Here, an increased degradation of the cell wall polysaccharides, mainly pectin, increases anthocyanin extractability from the tissue. Concurrently, cell wall breakage also results in the liberation of polysaccharides, which might beneficially affect anthocyanin stability by forming protective complexes. However, detailed research on controlled cell wall polysaccharide degradation, the characterization of the resultant polysaccharides, and related molecular complexation properties are lacking so far. The present thesis addresses these essential aspects contributing to the understanding of molecular processes that take place during red juice production.

The novel technology of ultrasound and particularly the applied ultrasound-assisted enzymatic maceration (UAEM) has gained great attention for gentle juice production due to the potential of increasing juice yield and polyphenol extraction while reducing temperature and time. Therefore, the present study evaluates the potential of UAEM on pectin degradation under reduced temperature and time in relation to the characteristics of the arising oligosaccharides and polysaccharides (**Chapter 2**). The aim of the present investigation was to achieve pectin degradation at least as effective as the commercial enzymatic treatment under conventional conditions.

The UAEM degradation mechanisms of cell wall polysaccharides are not clear so far. While previous studies postulated an enhanced pectin degradation by a synergistic effect of the pectinolytic activity of polygalacturonase (PG) under ultrasound treatment, the present study additionally analyzes all other pectinolytic enzyme activities of the applied enzyme preparation for individual synergistic effects by ultrasonication.

Soluble pectic polysaccharides are assumed to enhance anthocyanin stability by protective complexation. Indeed, interactions between anthocyanins and polysaccharides are multifaceted depending on the structural properties of both counterparts. During juice maceration, diverse pectic polymers are generated and modified that have not been considered in mechanisms of anthocyanin-complexation so far. Therefore, the present study aims to extend the knowledge of beneficial structural characteristics of both counterparts for protective complexations. For this purpose, interaction experiments in the present study were conducted with different pectic polysaccharides that were first modified by enzymatic and ultrasound treatments (EMP and UMP, respectively), thereby mimicking maceration. After this, the resultant polysaccharides were incubated with different anthocyanins (**Chapter 3**). Since anthocyanin-pectin complexation mechanisms have been distinguished between a fast, direct binding and slower anthocyanin stacking, complexation experiments have to be considered for both short and long-time incubations, evaluating the binding mechanisms and protective properties under both the conditions.

In summary, the aim of this thesis was to study the molecular mechanisms of polysaccharide degradation, in particular pectin leading to soluble polymers that possess complexation properties toward anthocyanins with protective effects.

The specific aims of the individual studies (**Chapters 2** and **3**) are:

- The evaluation of an UAEM continuous flow system at reduced temperature of 30 °C considering the degrading efficiency of the common enzymatic batch maceration at 50 °C (**Chapter 2**)
- The characterization of the resultant set of polysaccharides and oligosaccharides after UAEM, ultrasound, and enzyme treatment, especially in view of smaller oligomers (<30 kDa) (**Chapter 2**)
- The examination of individual synergistic effects by ultrasound application towards the main pectinolytic enzymes, that is, polygalacturonase (PG), pectin lyase (PL) and, pectin methylesterase (PME) (**Chapter 2**)
- The evaluation of structural dependencies in the interaction between anthocyanins and pectin using anthocyanins varying in their aglycone or glycoside and structurally different pectic polysaccharides (**Chapter 3**)
- The consideration of juice production-induced effects on anthocyanin-pectin-complexation and complex properties by pectin modifications, using native and maceration-like modified pectin (EMP and UMP) (**Chapter 3**)
- Examining the stability and binding period of complexations by comparing interaction experiments after two hours of incubation and two weeks of storage (**Chapter 3**)

References

Acosta-Estrada, B. A., Gutiérrez-Uribe, J. A., & Serna-Saldívar, S. O. (2014). Bound phenolics in foods, a review. *Food Chemistry, 152*, 46–55.

Andersen, Ø. M., & Jordheim, M. (2005). 10 The anthocyanins, In: Andersen, Ø. M. & Markham, K. R., (Eds.). *Flavonoids: Chemistry, biochemistry and applications*. CRC Press.

Arts, I. C., & Hollman, P. C. (2005). Polyphenols and disease risk in epidemiologic studies. *American Journal of Clinical Nutrition*, *81*(1), 317S–325S.

Ayestarán, B., Guadalupe, Z., & León, D. (2004). Quantification of major grape polysaccharides (*Tempranillo v.*) released by maceration enzymes during the fermentation process. *Analytica Chimica Acta*, *513*(1), 29–39.

Bagger-Jørgensen, R., & Meyer, A. S. (2004). Effects of different enzymatic pre-press maceration treatments on the release of phenols into blackcurrant juice. *European Food Research and Technology*, *219*(6), 620–629.

Birt, D. F., & Jeffery, E. (2013). Flavonoids. *Advances in Nutrition*, *4*(5), 576–577.

Brownmiller, C., Howard, L. R., & Prior, R. L. (2008). Processing and storage effects on monomeric anthocyanins, percent polymeric color, and antioxidant capacity of processed blueberry products. *Journal of Food Science*, *73*(5), H72–H79.

Buchert, J., Koponen, J. M., Suutarinen, M., Mustranta, A., Lilie, M., Törrönen, R., & Poutanen, K. (2005). Effect of enzyme-aided pressing on anthocyanin yield and profiles in bilberry and blackcurrant juices. *Journal of the Science of Food and Agriculture*, *85*(15), 2548–2556.

Buchweitz, M., Speth, M., Kammerer, D. R., & Carle, R. (2013a). Impact of pectin type on the storage stability of black currant (*Ribes nigrum* L.) anthocyanins in pectic model solutions. *Food Chemistry*, *139*(1–4), 1168–1178.

Buchweitz, M., Speth, M., Kammerer, D. R., & Carle, R. (2013b). Stabilisation of strawberry (*Fragaria x ananassa* Duch.) anthocyanins by different pectins. *Food Chemistry*, *141*(3), 2998–3006.

Carrera, C., Ruiz-Rodríguez, A., Palma, M., & Barroso, C. G. (2012). Ultrasound assisted extraction of phenolic compounds from grapes. *Analytica Chimica Acta*, *732*, 100–104.

Castaneda-Ovando, A., de Lourdes Pacheco-Hernandez, M., Paez-Hernandez, M. E., Rodriguez, J. A., & Galan-Vidal, C. A. (2009). Chemical studies of anthocyanins: A review. *Food Chemistry*, *113*(4), 859–871.

Chanliaud, E., & Gidley, M. J. (1999). *In vitro* synthesis and properties of pectin/ *Acetobacter xylinus* cellulose composites. *The Plant Journal*, *20*(1), 25–35.

Cheynier, V., Comte, G., Davies, K., Lattanzio, V., & Martens, S. (2013). Plant phenolics: Recent advances on their biosynthesis, genetics, and ecophysiology. *Plant Physiology and Biochemistry : PPB / Societe Francaise de Physiologie Vegetale*, *72*, 1–20.

Clifford, M. N. (2000). Anthocyanins – nature, occurence and dietary burden. *Journal of the Science of Food and Agriculture*, *80*, 1118–1125.

Cosgrove, D. J. (2005). Growth of the plant cell wall. *Nature Reviews Molecular Cell Biology*, *6*(11), 850–861.

Cui, R., & Zhu, F. (2021). Ultrasound modified polysaccharides: A review of structure, physicochemical properties, biological activities and food applications. *Trends in Food Science and Technology,* 107, 491–508.

Dalagnol, L. M. G., Dal Magro, L., Silveira, V. C. C., Rodrigues, E., Manfroi, V., & Rodrigues, R. C. (2017). Combination of ultrasound, enzymes and mechanical stirring: A new method to improve *Vitis vinifera* Cabernet Sauvignon must yield, quality and bioactive compounds. *Food and Bioproducts Processing*, *105*, 197–204.

de Vries, R. P., & Visser, J. (2001). *Aspergillus* enzymes involved in degradation of plant cell wall polysaccharides. *Microbiology and Molecular Biology Reviews*, *65*(4), 497–522.

Devore, E. E., Kang, J. H., Breteler, M. M. B., & Grodstein, F. (2012). Dietary intakes of berries and flavonoids in relation to cognitive decline. *Annals of Neurology*, 72(1), 135-143.

Doco, T., & Brillouet, J. M. (1993). Isolation and characterisation of a rhamnogalacturonan II from red wine. *Carbohydrate Research*, *243*(2), 333–343.

Doco, T., Williams, P., Vidal, S., & Pellerin, P. (1997). Rhamnogalacturonan II, a dominant polysaccharide in juices produced by enzymic liquefaction of fruits and vegetables. *Carbohydrate Research*, *297*(2), 181–186.

Ducasse, M. A., Williams, P., Canal-Llauberes, R. M., Mazerolles, G., Cheynier, V., & Doco, T. (2011). Effect of macerating enzymes on the oligosaccharide profiles of merlot red wines. *Journal of Agricultural and Food Chemistry*, *59*(12), 6558–6567.

European Union (2001). COUNCIL DIRECTIVE 2001/112/EC of 20 December 2001 *relating to fruit juices and certain similar products intended for human consumption*. Official Journal of the European Communities § (2001). Retrieved from http://extwprlegs1.fao.org/docs/pdf/eur34844.pdf

Feng, H., Barbosa-Cánovas, G. V., & Weiss, J. (2011). *Ultrasound Technologies for Food and Bioprocessing*. (Vol 1, p.599). New York, NY: Springer.

Fernandes, A., Brás, N. F., Mateus, N., & de Freitas, V. (2015). Correction to "Understanding the Molecular Mechanism of Anthocyanin Binding to Pectin." *Langmuir, 31*(5), 1866–1866.

Giusti, M. M., & Wrolstad, R. E. (2003). Acylated anthocyanins from edible sources and their applications in food systems. *Biochemical Engineering Journal*, *14*(3), 217–225.

DGE - Deutsche Gesellschaft für Ernährung e.V. (2007). Boeing, H., Bechthold, A., Bub, A., Ellinger, S., Haller, D., Kroke, A., ... Watzl, B. *Stellungnahme für Obst und Gemüse in der Prävention chronischer Krankheiten.* Bonn. Retrieved from https://www.dge.de/fileadmin/public/doc/ws/stellungnahme/Stellungnahme-OuG-Praevention-chronischer-Krankheiten-2007-09-29.pdf

GMF - Vereinigung Getreide-, Markt- und Ernährungsforschung (2002). Becker, H.-G. *Ballaststoffe in unseren Lebensmitteln*. Detmold. Retrieved from https://www.gmf-info.de/ballaststoffe.pdf

Grant, G. T., Morris, E. R., Rees, D. A., Smith, P. J. C., & Thom, D. (1973). Biological interactions between polysaccharides and divalent cations: The egg-box model. *FEBS Letters*, *32*(1), 195–198.

Grassin, C., & Coutel, Y. (2010). Enzymes in fruit and vegetable processing and juice extraction. In: R. J. Whitehurst & M. Van Oort, (Eds.). *Enzymes in Food Technology* (2nd edition, p. 372). Wiley-Blackwell.

Guadalupe, Z., & Ayestarán, B. (2007). Polysaccharide profile and content during the vinification and aging of Tempranillo red wines. *Journal of Agricultural and Food Chemistry*, *55*(26), 10720–10728.

Hager, J. T., Howard, L. R., & Prior, R. L. (2008). Processing and storage effects on monomeric anthocyanins, percent polymeric color, and antioxidant capacity of processed blackberry products. *Journal of Agricultural and Food Chemistry*, *56*(3), 689–695.

Harholt, J., Suttangkakul, A., & Vibe Scheller, H. (2010). Biosynthesis of Pectin. *Plant Physiology*, *153*(2), 384–395.

Heffels, P., Bührle, F., Schieber, A., & Weber, F. (2017). Influence of common and excessive enzymatic treatment on juice yield and anthocyanin content and profile during bilberry (*Vaccinium myrtillus* L.) juice production. *European Food Research and Technology*, *243*(1), 59–68.

Heffels, P., Weber, F., & Schieber, A. (2015). Influence of accelerated solvent extraction and ultrasound-assisted extraction on the anthocyanin profile of different *Vaccinium* species in the context of statistical models for authentication. *Journal of Agricultural and Food Chemistry*, *63*(34), 7532–7538.

Hilz, H., Bakx, E. J., Schols, H. A., & Voragen, A. (2005). Cell wall polysaccharides in black currants and bilberries - Characterisation in berries, juice, and press cake. *Carbohydrate Polymers*, *59*(4), 477–488.

Hilz, H., Lille, M. L., Poutanen, K. P., Schols, H., & Voragen, A. (2006). Combined enzymatic and high-pressure processing affect cell wall polysaccharides in berries. *Journal of Agricultural and Food Chemistry*, 54(4), 1322–1328.

Hilz, H., Williams, P., Doco, T., Schols, H. A., & Voragen, A. (2006). The pectic polysaccharide rhamnogalacturonan II is present as a dimer in pectic populations of bilberries and black currants in muro and in juice. *Carbohydrate Polymers*, *65*(4), 521–528.

Holzwarth, M., Korhummel, S., Carle, R., & Kammerer, D. R. (2012). Impact of enzymatic mash maceration and storage on anthocyanin and color retention of pasteurized strawberry purées. *European Food Research and Technology*, *234*(2), 207–222.

Holzwarth, M., Korhummel, S., Siekmann, T., Carle, R., & Kammerer, D. R. (2013). Influence of different pectins, process and storage conditions on anthocyanin and colour retention in strawberry jams and spreads. *LWT - Food Science and Technology*, *52*(2), 131–138.

Jakobek, L. (2015). Interactions of polyphenols with carbohydrates, lipids and proteins. *Food Chemistry*, *175*, 556–567.

Jakobek, L., Šeruga, M., Medvidović-Kosanović, M., & Novak, I. (2007). Anthocyanin content and antioxidant activity of various red fruit juices. *Deutsche Lebensmittel-Rundschau*, *103*(2), 58–64.

Jones-Moore, H. R., Jelley, R. E., Marangon, M., & Fedrizzi, B. (2021). The polysaccharides of winemaking: From grape to wine. *Trends in Food Science & Technology*, *111*, 731–740.

Kentish, S., & Feng, H. (2014). Applications of power ultrasound in food processing. *Annual Review of Food Science and Technology*, *5*, 263–284.

Knorr, D., Froehling, A., Jaeger, H., Reineke, K., Schlueter, O., & Schoessler, K. (2011). Emerging technologies in food processing. *Annual Review of Food Science and Technology*, *2*(1), 203–235.

Kopjar, M., Piližota, V., Tiban, N. N., Šubarić, D., Babić, J., Ačkar, D., & Sajdl, M. (2009). Strawberry jams: Influence of different pectins on colour and textural properties. *Czech Journal of Food Sciences*, *27*(1), 20–28.

Koponen, J. M., Buchert, J., Poutanen, K. S., & Törrönen, A. R. (2008). Effect of pectinolytic juice production on the extractability and fate of bilberry and black currant anthocyanins. *European Food Research and Technology*, *227*(2), 485–494.

Kressmann, R. (2001). *The chemical composition of soluble colloids in the juice of black currants and their effect on processing technology*. Ph.D. thesis (in German), Justus-Liebig University of Gießen, GER.

Laaksonen, O., Sandell, M., Nordlund, E., Heiniö, R. L., Malinen, H. L., Jaakkola, M., & Kallio, H. (2012). The effect of enzymatic treatment on blackcurrant (*Ribes nigrum*) juice flavour and its stability. *Food Chemistry*, *130*(1), 31–41.

Landbo, A.-K., & Meyer, A. S. (2004). Effects of different enzymatic maceration treatments on enhancement of anthocyanins and other phenolics in black currant juice. *Innovative Food Science & Emerging Technologies*, *5*(4), 503–513.

Landbo, A. K., & Meyer, A. S. (2001). Enzyme-assisted extraction of antioxidative phenols from black currant juice press residues (*Ribes nigrum*). *Journal of Agricultural and Food Chemistry*, *49*(7), 3169–3177.

Le Bourvellec, C., Bouchet, B., & Renard, C. (2005). Non-covalent interaction between procyanidins and apple cell wall material. Part III. Study on model polysaccharides. *Biochimica et Biophysica Acta (BBA)-General Sujects*, *1725*(1), 10–18.

Lee, J., Durst, R. W., & Wrolstad, R. E. (2002). Impact of juice processing on blueberry anthocyanins and polyphenolics: Comparison of two pretreatments. *Journal of Food Science*, *67*(5), 1660–1667.

Leong, S. Y., & Oey, I. (2017). Berry Juices. In: I. Aguiló-Aguayo, L. Plaza, & J. Wiley, (Eds.). *Innovative Technologies in Beverage Processing*. Chichester, UK: John Wiley & Sons, Ltd.

Li, F., Chen, G., Zhang, B., & Fu, X. (2017). Current applications and new opportunities for the thermal and non-thermal processing technologies to generate berry product or extracts with high nutraceutical contents. *Food Research International*, *100*, 19–30.

Lieu, L. N., & Le, V. V. M. (2010). Application of ultrasound in grape mash treatment in juice processing. *Ultrasonics Sonochemistry*, *17*(1), 273–279.

Liu, X., Le Bourvellec, C., & Renard, C. M. G. C. (2020). Interactions between cell wall polysaccharides and polyphenols: Effect of molecular internal structure. *Comprehensive Reviews in Food Science and Food Safety*, *19*(6), 3574–3617.

Liu, Y., Ying, D., Sanguansri, L., & Augustin, M. A. (2019). Comparison of the adsorption behaviour of catechin onto cellulose and pectin. *Food Chemistry*, *271*, 733–738.

Ma, X., Cai, J., & Liu, D. (2020). Ultrasound for pectinase modification: An investigation into potential mechanisms. *Journal of the Science of Food and Agriculture*, *100*(12), 4636–4642.

Ma, X., Zhang, L., Wang, W., Zou, M., Ding, T., Ye, X., & Liu, D. (2016). Synergistic effect and mechanisms of combining ultrasound and pectinase on pectin hydrolysis. *Food and Bioprocess Technology*, *9*(7), 1249–1257.

Maier, T., Fromm, M., Schieber, A., Kammerer, D. R., & Carle, R. (2009). Process and storage stability of anthocyanins and non-anthocyanin phenolics in pectin and gelatin gels enriched with grape pomace extracts. *European Food Research and Technology*, *229*(6), 949–960.

Mazza, G., & Brouillard, R. (1987). Recent developments in the stabilization of anthocyanins in food products. *Food Chemistry*, *25*(3), 207–225.

Mazzaracchio, P., Pifferi, P., Kindt, M., Munyaneza, A., & Barbiroli, G. (2004). Interactions between anthocyanins and organic food molecules in model systems. *International Journal of Food Science and Technology*, *39*(1), 53–59.

McCann, M. C., Wells, B., & Roberts, K. (1990). Direct visualization of cross-links in the primary plant cell wall. *Journal of Cell Science*, *96*(2), 323–334.

McManus, J. P., Davis, K. G., Beart, J. E., Gaffney, S. H., Lilley, T. H., & Haslam, E. (1985). Polyphenol interactions. Part 1. Introduction; some observations on the reversible complexation of polyphenols with proteins and polysaccharides. *Journal of the Chemical Society, Perkin Transactions 2*, *28*(2), 1429.

Medina-Plaza, C., Beaver, J. W., Miller, K. V., Lerno, L., Dokoozlian, N., Ponangi, R., … Oberholster, A. (2020). Cell wall-anthocyanin interactions during red wine fermentation-like conditions. *American Journal of Enology and Viticulture*, *71*(2), 149–156.

Mieszczakowska-Frac, M., Markowski, J., Zbrzezniak, M., & Plocharski, W. (2012). Impact of enzyme on quality of blackcurrant and plum juices. *LWT - Food Science and Technology*, *49*(2), 251–256.

Molan, A. L., Lila, M. A., Mawson, J., & De, S. (2009). *In vitro* and *in vivo* evaluation of the prebiotic activity of water-soluble blueberry extracts. *World Journal of Microbiology and Biotechnology*, *25*(7), 1243–1249.

Moore, M. A., Park, C. B., & Tsuda, H. (1998). Soluble and insoluble fiber influences on cancer development. *Critical Reviews in Oncology/Hematology*, *27*(3), 229–242.

Nadar, S. S., & Rathod, V. K. (2017). Ultrasound assisted intensification of enzyme activity and its properties: A mini-review. *World Journal of Microbiology and Biotechnology*, *33*(9), 170.

Nguyen, V. P. T., Le, T. T., & Le, V. V. M. (2013). Application of combined ultrasound and cellulase preparation to guava (*Psidium guajava*) mash treatment in juice processing: Optimization of biocatalytic conditions by response surface methodology. *International Food Research Journal*, *20*(1), 377–381.

O'Neill, M. A., Warrenfeltz, D., Kates, K., Pellerin, P., Doco, T., Darvill, A. G., & Albersheim, P. (1996). Rhamnogalacturonan-II, a pectic polysaccharide in the walls of growing plant cell, forms a dimer that is covalently cross-linked by a borate ester. *In vitro* conditions for the formation and hydrolysis of the dimer. *Journal of Biological Chemistry, 271*(37), 22923–22930.

Ochoa-Villarreal, M., Aispuro-Hernndez, E., Vargas-Arispuro, I., & Martinez-Téllez, M.A. (2012). Plant cell wall polymers: Function, structure and biological activity of their derivatives. In: A. De Souza Gomes, (Eds.). *Polymerization* (4th edition). Rijeka, CRO: InTech.

Odriozola-Serrano, I., Soliva-Fortuny, R., & Martín-Belloso, O. (2008). Phenolic acids, flavonoids, vitamin C and antioxidant capacity of strawberry juices processed by high-intensity pulsed electric fields or heat treatments. *European Food Research and Technology, 228*(2), 239–248.

Ortega-Regules, A., Romero-Cascales, I., Ros-García, J. M., López-Roca, J. M., & Gómez-Plaza, E. (2006). A first approach towards the relationship between grape skin cell-wall composition and anthocyanin extractability. *Analytica Chimica Acta, 563*(1-2), 26–32.

Ortega-Regules, Ana, Ros-García, J. M., Bautista-Ortín, A. B., López-Roca, J. M., & Gómez-Plaza, E. (2008). Differences in morphology and composition of skin and pulp cell walls from grapes (*Vitis vinifera* L.): Technological implications. *European Food Research and Technology, 227*(1), 223–231.

Padayachee, A., Netzel, G., Netzel, M., Day, L., Zabaras, D., Mikkelsen, D., & Gidley, M. J. (2012). Binding of polyphenols to plant cell wall analogues - Part 1: Anthocyanins. *Food Chemistry, 134*(1), 155–161.

Pandey, K. B., & Rizvi, S. I. (2009). Plant polyphenols as dietary antioxidants in human health and disease. *Oxidative Medicine and Cellular Longevity*, 2(5), 270–278.

Paniwnyk, L. (2017). Applications of ultrasound in processing of liquid foods: A review. *Ultrasonics Sonochemistry, 38*, 794–806.

Pappas, E., & Schaich, K. M. (2009). Phytochemicals of cranberries and cranberry products: Characterization, potential health effects, and processing stability. *Critical Reviews in Food Science and Nutrition, 49*(9), 741–781.

Patras, A., Brunton, N. P., O'Donnell, C., & Tiwari, B. K. (2010). Effect of thermal processing on anthocyanin stability in foods; mechanisms and kinetics of degradation. *Trends in Food Science and Technology, 21*(1), 3–11.

Pauly, M., Albersheim, P., Darvill, A., & York, W. S. (1999). Molecular domains of the cellulose/xyloglucan network in the cell walls of higher plants. *The Plant Journal*, *20*(6), 629–639.

Pellerin, P., Doco, T., Vidal, S., Williams, P., Brillouet, J. M., & O'Neill, M. A. (1996). Structural characterization of red wine rhamnogalacturonan II. *Carbohydrate Research*, *290*(2), 183–197.

Phan, A. D. T., Flanagan, B. M., D'Arcy, B. R., & Gidley, M. J. (2017). Binding selectivity of dietary polyphenols to different plant cell wall components: Quantification and mechanism. *Food Chemistry*, *233*, 216–227.

Pinelo, M., Arnous, A., & Meyer, A. S. (2006). Upgrading of grape skins: Significance of plant cell-wall structural components and extraction techniques for phenol release. *Trends in Food Science and Technology*, *17*(11), 579–590.

Poiana, M.-A., Munteanu, M.-F., Bordean, D.-M., Gligor, R., & Alexa, E. (2013). Assessing the effects of different pectins addition on color quality and antioxidant properties of blackberry jam. *Chemistry Central Journal*, *7*(1), 121.

Puupponen-Pimiä, R., Nohynek, L., Alakomi, H. L., & Oksman-Caldentey, K. M. (2005). The action of berry phenolics against human intestinal pathogens. *BioFactors*, 23(4), 234–251.

Rein, M. J. (2005). *Copigmentation reactions and color stability of berry anthocyanins*. Ph.D. thesis, Universtiy of Helsinki.

Rein, M. J., & Heinonen, M. (2004). Stability and Enhancement of Berry Juice Color. *Journal of Agricultural and Food Chemistry*, *52*(10), 3106–3114.

Renard, C. M. G. C., Baron, A., Guyot, S., & Drilleau, J. F. (2001). Interactions between apple cell walls and native apple polyphenols: Quantification and some consequences. *International Journal of Biological Macromolecules*, *29*(2), 115–125.

Rinaldi, M., Caligiani, A., Borgese, R., Palla, G., Barbanti, D., & Massini, R. (2013). The effect of fruit processing and enzymatic treatments on pomegranate juice composition, antioxidant activity and polyphenols content. *LWT - Food Science and Technology*, *53*(1), 355–359.

Ropartz, D., & Ralet, M.-C. (2020). 2 Pectin Structure. In: V. Kontogiorgos, (Eds.). *Pectin: Technological and Physiological Properties*. Springer Nature Switzerland AG.

Rossi, M., Giussani, E., Morelli, R., Lo Scalzo, R., Nanic, R. C., & Torreggiani, D. (2003). Effect of fruit blanching on phenolics and radical scavenging activity of highbush blueberry juice. *Food Research International*, *36*(9–10), 999–1005.

Sadilova, E., Stintzing, F. C., Kammerer, D. R., & Carle, R. (2009). Matrix dependent impact of sugar and ascorbic acid addition on color and anthocyanin stability of black carrot, elderberry and strawberry single strength and from concentrate juices upon thermal treatment. *Food Research International*, *42*(8), 1023–1033.

Scalbert, A., Manach, C., Morand, C., Rémésy, C., & Jiménez, L. (2005). Dietary polyphenols and the prevention of diseases. *Critical Reviews in Food Science and Nutrition*, *45*(4), 287–306.

Scheller, H. V., & Ulvskov, P. (2010). Hemicelluloses. *Annual Review of Plant Biology*, *61*(1), 263–289.

Schols, H. A., & Voragen, A. G. J. (1996). Complex pectins: Structure elucidation using enzymes. In: J. Visser J., A. G. J. Voragen, (Eds.). *Progress in Biotechnology (Vol. 14, pp. 3-19)*. Amsterdam, NED: Elsevier Science.

Schols, Henk A. (1995). *Structural characterization of pectic hairy regions isolated from apple cell walls*. Ph.D. thesis, Wageningen Agricultural University, NED.

Shikov, V., Kammerer, D. R., Mihalev, K., Mollov, P., & Carle, R. (2008). Heat stability of strawberry anthocyanins in model solutions containing natural copigments extracted from rose (*Rosa damascena* Mill.) petals. *Journal of Agricultural and Food Chemistry, 56*(18), 8521–8526.

Skrede, G., Wrolstad, R. E., & Durst, R. W. (2000). Changes in anthocyanins and polyphenolics during juice processing of highbush blueberries (*Vaccinium corymbosum* L.). *Journal of Food Science*, *65*(2), 357–364.

Spencer, J. P. E., Abd El Mohsen, M. M., Minihane, A.-M., & Mathers, J. C. (2008). Biomarkers of the intake of dietary polyphenols: Strengths, limitations and application in nutrition research. *British Journal of Nutrition*, *99*(1), 12–22.

Srivastava, A., Akoh, C. C., Yi, W., Fischer, J., & Krewer, G. (2007). Effect of storage conditions on the biological activity of phenolic compounds of blueberry extract packed in glass bottles. *Journal of Agricultural and Food Chemistry, 55*(7), 2705–2713 .

Stow, J. (1993). Effect of calcium ions on apple fruit softening during storage and ripening. *Postharvest Biology and Technology*, *3*(1), 1–9.

Tchabo, W., Ma, Y., Engmann, F. N., & Zhang, H. (2015). Ultrasound-assisted enzymatic extraction (UAEE) of phytochemical compounds from mulberry (*Morus nigra*) must and optimization study using response surface methodology. *Industrial Crops and Products*, *63*, 214–225.

Trouillas, P., Sancho-García, J. C., de Freitas, V., Gierschner, J., Otyepka, M., & Dangles, O. (2016). Stabilizing and modulating color by copigmentation: Insights from theory and experiment. *Chemical Reviews*, *116*(9), 4937–4982.

VdF - Verband der deutschen Fruchtsaft-Industrie e.V. (2019). *Daten und Fakten der deutschen Fruchtsaft-Industrie*. Bonn. Retrieved from www.fruchtsaft.de

Vidal, S., Francis, L., Williams, P., Kwiatkowski, M., Gawel, R., Cheynier, V., & Waters, E. (2004). The mouth-feel properties of polysaccharides and anthocyanins in a wine like medium. *Food Chemistry*, *85*(4), 519–525.

Vidal, S., Williams, P., O'Neill, M. A., & Pellerin, P. (2001). Polysaccharides from grape berry cell walls. Part I. tissue distribution and structural characterization of the pectic polysaccharids. *Carbohydrate Polymers*, *45*(4), 315–323.

Vierhuis, E., York, W. S., Kolli, V. S. K., Vincken, J.-P., Schols, H. A., Van Alebeek, G.-J. W., & Voragen, A. G. J. (2001). Structural analyses of two arabinose containing oligosaccharides derived from olive fruit xyloglucan: XXSG and XLSG. *Carbohydrate Research*, *332*(3), 285–297.

Vilkhu, K., Mawson, R., Simons, L., & Bates, D. (2008). Applications and opportunities for ultrasound assisted extraction in the food industry – A review. *Innovative Food Science and Emerging Technologies*, *9*(2), 161–169.

Voragen, A. G. J., Coenen, G. J., Verhoef, R. P., & Schols, H. A. (2009). Pectin, a versatile polysaccharide present in plant cell walls. *Structural Chemistry*, *20*(2), 263–275.

Wang, D., Yan, L., Ma, X., Wang, W., Zou, M., Zhong, J., ... Liu, D. (2018). Ultrasound promotes enzymatic reactions by acting on different targets: Enzymes, substrates and enzymatic reaction systems. *International Journal of Biological Macromolecules*, *119*, 453–461.

Wang, W., Chen, W., Zou, M., Lv, R., Wang, D., Hou, F., ... Liu, D. (2018). Applications of power ultrasound in oriented modification and degradation of pectin: A review. *Journal of Food Engineering 234*, 98–107.

Weber, F., & Larsen, L. R. (2017). Influence of fruit juice processing on anthocyanin stability. *Food Research International*, *100*, 354–365.

Whitaker, J. R., Voragen, A. G. J., & Wong, D. W. S., (Eds.). (2003). *Handbook of food enzymology*. New Yor, NY: Marcel Dekker.

Whitcombe, A. J., O'Neill, M. A., Steffan, W., Albersheim, P., & Darvill, A. G. (1995). Structural characterization of the pectic polysaccharide, rhamnogalacturonan-II. *Carbohydrate Research*, *271*(1), 15–29.

Wilkes, K., Howard, L. R., Brownmiller, C., & Prior, R. L. (2014). Changes in chokeberry (*Aronia melanocarpa* L.) polyphenols during juice processing and storage. *Journal of Agricultural and Food Chemistry*, *62*(18), 4018–4025.

Wrolstad, R. E., Durst, R. W., & Lee, J. (2005). Tracking color and pigment changes in anthocyanin products. *Trends in Food Science & Technology*, *16*(9), 423–428.

Wu, X., & Prior, R. L. (2005). Systematic identification and characterization of anthocyanins by HPLC-ESI-MS/MS in common foods in the United States: Fruits and berries. *Journal of Agricultural and Food Chemistry*, *53*(7), 2589–2599.

Yan, J. K., Wang, Y. Y., Ma, H. Le, & Wang, Z. Bin. (2016). Ultrasonic effects on the degradation kinetics, preliminary characterization and antioxidant activities of polysaccharides from *Phellinus linteus* mycelia. *Ultrasonics Sonochemistry*, *29*(June), 251–257.

Zhao, C.-L., Yu, Y.-Q., Chen, Z.-J., Wen, G.-S., Wei, F.-G., Zheng, Q., ... Xiao, X.-L. (2017). Stability-increasing effects of anthocyanin glycosyl acylation. *Food Chemistry*, *214*, 119–128.

Zinoviadou, K. G., Galanakis, C. M., Brnčić, M., Grimi, N., Boussetta, N., Mota, M. J., ... Barba, F. J. (2015). Fruit juice sonication: Implications on food safety and physicochemical and nutritional properties. *Food Research International*, *77*, 743–752.

Chapter 2

Effects of Ultrasound on the Enzymatic Degradation of Pectin

Ultrasound-assisted enzymatic maceration (UAEM) has gained considerable interest in the fruit juice industry, owing to its potential to increase juice yield and content of polyphenols while simultaneously saving time and energy. In this study, the effects of UAEM (ultrasonic probe, 20 kHz, 21 W*cm^{-2} and 33 W*cm^{-2}) on pectin degradation in a continuous circulation system were investigated over 60 and 90 min. Main pectinolytic enzymes activities (polygalacturonase, pectin lyase and pectin methylesterase) of a commercial enzyme preparation were examined for individual synergistic effects with US. Pectin hydrolysis by UAEM differed significantly compared to treatment with ultrasound or enzymes alone regarding the profile of degradation products compared to treatment with ultrasound or enzymes alone. Ultrasound fragmented pectin to less branched oligomers of medium molecular weight (Mp approx. 150 kDa), which were further degraded by pectinolytic activities. The low molecular weight fraction (<30 kDa), which is known to be beneficial for juice-quality by adding nutritional value and stabilizing polyphenols, was enriched in small oligomers of homogalacturonan-derived, rhamnogalacturonan I-derived, and rhamnogalacturonan II-derived residues. Synergistic effects of ultrasound application enhanced the effective activities of polygalacturonase and pectin lyase and even prolonged their performance over 90 min, whereas the effective activity of pectin methylesterase was not affected. Final marker concentrations determined by each enzyme assay revealed a considerable higher total process output after UAEM treatment at reduced temperature (30 °C) comparable to the output after conventional batch maceration at 50 °C. The obtained results demonstrate the high potential of UAEM to produce high-quality juice by controlling pectin degradation while reducing process temperature and equally highlight the matrix and enzyme specific effects of a simultaneous US treatment.

Keywords: ultrasound-assisted enzymatic maceration, pectin degradation, low molecular weight oligomers, polygalacturonase, pectin lyase, pectin methylesterase, synergistic effects

Abbreviations: DA, degree of acetylation; DM, degree of methylation; GalAc, galacturonic acid; HG, homogalacturonan; HPSEC, high-performance size exclusion chromatography; MW, molecular weight; Mp, peak molecular weight; PG, polygalacturonase; PL, pectin lyase; PME, pectin methylesterase; RG I, rhamnogalacturonan I; RG II, rhamnogalacturonan II; UAEM, ultrasound-assisted enzymatic maceration; US, ultrasound

This chapter has been published:

Larsen, L. R., van der Weem, J., Caspers-Weiffenbach, R., Schieber, A., & Weber, F. (2021) Effects of Ultrasound on the Enzymatic Degradation of Pectin. *Ultrasonics Sonochemistry*, (72), 105465

1 Introduction

Ultrasound (US) technology has gained increasing attention in food processing in the past decade. Particularly in fruit juice production, this emerging technology provides several promising advantages like minimizing processing time, improving sensory quality, and ensuring microbial safety [1–3]. The use of high-intensity US offers a non-thermal alternative to conventional pasteurization techniques that avoids undesired side effects [4]. Benefits have recently been shown also for low-intensity US treatment, e.g., the enhanced yet milder extraction of natural compounds like polysaccharides or polyphenols with higher yields and reduced processing time [3,5]. Additionally, US treatment increases enzyme activities by synergistic effects, also resulting in higher extraction yields of plant materials compared to common enzymatic extractions [6,7].

In juice production, maceration plays a crucial role in defining the quality of the juice. In this process step, the extent of cell wall degradation determines the juice yield and extraction of natural compounds. The addition of pectinolytic enzyme during mashing primarily to increase juice yield is common industrial practice, especially in berry juice production due to the higher pectin content. Commercial enzyme preparations usually contain a mixture of pectinase activities, encompassing mainly polygalacturonase (PG), pectin lyase (PL), and pectin methylesterase (PME) accompanied by a wide range of minor side activities. These mixtures are therefore capable of hydrolyzing pectin being part of the complex polysaccharides of the plant cell wall into simpler molecules, thereby enhancing the release of phytochemicals like anthocyanins into the juice. Other polysaccharides like hemicellulose and cellulose remain more or less unchanged depending on the present side activities. [8–10] The quality and extent of cell wall degradation mainly depends on the activity and dosage of the applied enzymes, the enzyme's optimal conditions, maceration duration, and the complexity of the fruit´s cell wall matrix [11,12].

Pectin, the most abundant polysaccharide presented in the plant cell wall [13], has been demonstrated to strongly interact with phenolic compounds like anthocyanins. The complexation due to weak bonds like hydrophobic interactions and hydrogen bonds depends on the structure of both pectin and anthocyanins and protects the latter against oxidation and other degradation pathways. The molecular weight (MW) of pectin-derived polysaccharides affects the solubility of the complexes formed, thus determine the nutritional value and color quality of the juice. High MW complexes are very likely removed during pressing, leading to a significant loss of bound phenolic compounds. [14–17] Small oligomers will pass the production process into the juice due to their molecular size and higher solubility enriching the juices in fiber content and stabilizing valuable anthocyanins by complexation. [10,15,18] Because these interactions have the potential to greatly improve nutritional and sensory properties, the control of the extent of cell wall degradation and the size of arising oligosaccharides and polysaccharides during maceration is of pivotal importance for the quality of the juice.

Ultrasound-assisted enzymatic maceration (UAEM) represents a new technological approach to further increase fruit juice quality. To date, only a few studies demonstrated some selected beneficial effects, like higher or similar juice yields at reduced processing temperatures or in less time compared to conventional enzyme treatment. Concomitantly, the extraction of phytochemicals like phenolic acids, flavonoids, and especially anthocyanins was improved, the latter leading to an intensified color. Most of these studies used ultrasonic bath systems and treated samples batch-wise in flasks only at laboratory scale [1,2,19,20]. However, the juice industry strongly preferers continuous systems for juice processing which even fewer studies investigated requiring a flow cell with an integrated US probe. These authors focused on the effect on anthocyanin content, color, and shelf life of several juices in the context of microbial counts reduction [21,22]. To the best of our knowledge, those studies lack a detailed characterization of the arising pool of polysaccharides that strongly interacts with other fruit compounds and determines juice quality.

The US principle of a facilitated extraction primarily by cell wall disruption caused by collapsing cavitational bubbles increases cell membrane permeability and leads to enhanced mass transfer. In comparison with high-intensity US (>100 kHz) treatment, which induces chemical reactions like the formation of hydroxyl radicals, low-intensity US (approx. 20 kHz) recommended for extraction generates mechanical, cavitational, and thermal forces that act on the cell wall [23]. The US-induced forces can also alter the molecular structure of enzymes and their corresponding substrates, affecting significantly their activity [7]. Few studies reported a synergistic effect applying of US and pectinases, in particular PG [24,25]. However, synergistic effect toward the other main pectinolytic enzyme activities PL and PME in fruit juice preparations has not yet been described so far.

All these effects and other general US-induced effects, like cavitation phenomena, pressure oscillation, local shear stress, and the formation of radicals have been demonstrated to be highly variable depending on type of US setup, treatment duration, intensity, and food matrix [4,26]. Therefore, it is necessary to evaluate the distinct effects on matrix compounds individually to explain the underlying mechanisms.

The present study focuses on the characterization of pectin degradation by UAEM at a reduced maceration temperature of 30 °C in a continuous circulation system. The results were compared to a benchmark degradation of pectin by the same enzyme preparation at 50 °C for 60 min, employing parameters commonly used in fruit juice processing. To reduce interfering effects of other matrix components that would hamper the investigation in a berry mash, a sugar beet pectin model solution (pH 3.5) was used. Sugar beet pectin has a more complex structure with more similarities to berry pectin compared to industrially extracted apple or citrus pectin [10,27]. It contains higher amounts of neutral sugars due to higher contents of the rhamnogalacturonan I (RG I) and rhamnogalacturonan II (RG II) branched subunits, the higher degree of acetylation (DA), and a medium degree of methylation (DM), compared to citrus and apple pectin which mainly consist of linear galacturonic acid chains known as homogalacturonan (HG). To determine the distinct degradation effects of US and enzyme, as well as the combined effects, pectin polysaccharides and oligosaccharides were characterized by the viscosity they induce, high-performance size exclusion chromatography (HPSEC), DM, DA, and the content of three specific monomers in pectin's complex structure. Galacturonic acid (GalAc), rhamnose, and fucose were quantified because of their location in distinct subunits of pectin. Besides the determination of the specific enzyme activities of the three main pectinolytic enzymes PG, PL, and PME under optimal conditions, the effective enzyme activities were examined regarding the UAEM process conditions to reveal synergistic effects. In this study, synergistic effects for PL and PME of a commercial enzyme preparation by the low-intensity US were evaluated for the first time. The final marker concentration of each enzyme assay was measured and used to assess the total process output of UAEM.

2 Materials and Methods

2.1 Materials

Ultrapure water was obtained from a PURELAB flex 2 water purification system (ELGA LabWater, Paris, France). Ethanol (99.7%) and acetic acid were purchased from VWR (Mannheim, Germany). Ethanol (HPLC grade) and citric acid monohydrate (≥95.5%) were obtained from Carl Roth GmbH & Co. KG (Karlsruhe, Germany). Methanol (HPLC grade) and sulphuric acid (95%) were from Th. Geyer (Renningen, Germany). Formic acid (99.9%), malondialdehyde tetrabutylammonium salt (≥97%), potassium sodium tartrate tetrahydrate (≥98%), and 3,5-dinitrosalicylic acid (≥98%) were obtained from Sigma-Aldrich (St. Louis, MO, USA). n-Propanol, propionic acid, 2-thiobarbituric acid and sodium azide were purchased from Merk (Darmstadt, Germany); sodium hydroxide was from Honeywell (Morris Plains, NJ, USA). D-(+)-GalAc monohydrate (99%) was obtained from Fluka (Munich, Germany), sodium nitrate (99%) was from Acros Organics (Geel, Belgium), and trisodium citrate dihydrate (≥98%) was from Alpha Aesar (Ward Hill, MA, USA). Sugar beet pectin Betapec RU 301, apple pectin Classic AU-L 036/18 (low methylated, DM 38%) and apple pectin Classic AU-L 022/17 (high methylated, DM 71%) were kindly provided by Herbstreith & Fox GmbH & Co. KG (Neuenbürg, Germany).

2.2 Ultrasound-assisted enzymatic maceration (UAEM) treatment of pectin model solution in a continuous circulation system (30 °C)

A continuous circulation system (**Figure 2-1**), comparable to that described by Wong *et al.* [22], was used to sonicate a pectin model solution with added enzyme preparation (**Figure 2-2**). The experimental setup consisted of a thermostatic water bath (1) to maintain the temperature at 30 ± 2 °C during the treatment. The model solution (600 ml) was placed in a beaker (2), from which the solution was continuously circulated at a constant flow rate of 7.3 ml*sec^{-1} (0.44 L*min^{-1}) by a peristaltic pump (3; Ismatec ecoline from Cole-Parameter GmbH, Wertheim, Germany) through a cylindrical flow cell (4; FC100L1K-1S from Hielscher Ultrasonics GmbH, Teltow, Germany) equipped with a US

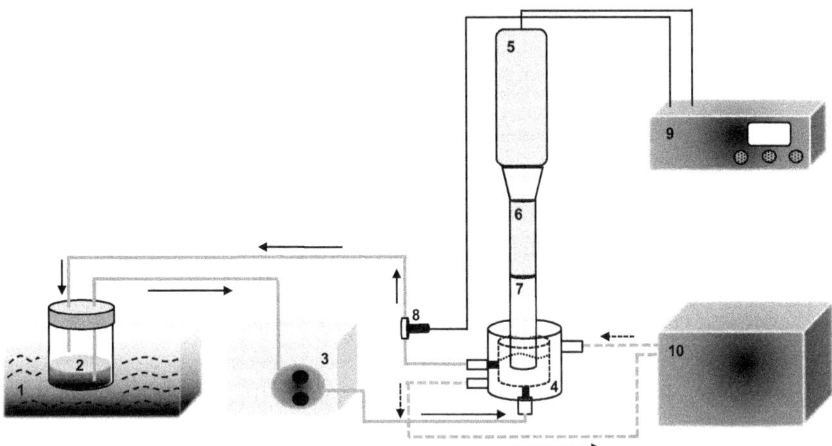

Figure 2-1: Schematic structure of continuous circulation system with ultrasound probe applied to sonicate pectin model solution; (1) thermostatic water bath, (2) sample beaker, (3) peristaltic pump (7.3 ml*sec^{-1}), (4) cylindrical flow-cell, (5) ultrasound processor, (6) booster horn, (7) ultrasound probe, (8) thermometer, (9) ultrasound generator, (10) cooling unit.

probe of 9 cm^2 (7). The flow cell underneath the probe had a volume of 9 ml. US treatments were performed by a US processor (5; UIP 1000hdT, 1000 W, 20 kHz, Hielscher, Teltow, Germany) fitted with a booster horn (6; 100% amplitude: 53 µm). The amplitude of the US generator (9) was set at 60% (UAEM 60%) or 90% (UAEM 90%), resulting in an energy input of 190 ± 3 W and 300 ± 6 W, respectively. Specific power density δ and specific intensity *I* were calculated by the following equations (1) - (3):

$$\delta_1 \left[\frac{W_{net}}{ml}\right] = \frac{P}{V_1} \qquad (1)$$

$$\delta_2 \left[\frac{W_{net}}{ml}\right] = \frac{P}{V_2} \qquad (2)$$

$$I \left[\frac{W_{net}}{cm^2}\right] = \frac{P}{A} \qquad (3), \text{ where}$$

δ... specific density (W*ml^{-1}), I... specific intensity (W*cm^{-2}), P... ultrasonic power (quantified by probe processor), V_1... volume of flow cell (9 ml), V_2... sample volume in system (600 ml), A... surface area of probe (cm^{-2})

The duration of the treatments varied between 60 and 90 min. The sample temperature was measured directly after sonication (8) and was controlled by a cooling unit (10; ProfiCool Novus, PCNO 30.03-NED from NationalLab GmbH, Moelln, Germany).

The applied pectin model solution was prepared by dissolving sugar beet pectin in a 0.05 M sodium citrate buffer (pH 3.5) at a concentration of 1% (w/v) by stirring the suspension overnight. The enzyme preparation (Rohapect®Classic, AB Enzymes GmbH, Darmstadt, Germany) is typically applied for fruit juice production during the mashing process. The model solution was tempered to 30 °C ± 2 °C before 100 ppm of the enzyme was added. Effects of US application alone were examined in the same system at highest amplitude (90%) without enzyme addition. Influences of the enzyme preparation alone were determined by running the continuous circulation system without US treatment (0% amplitude). All experiments were carried out in duplicate. Aliquots of samples were heated (100 °C, 3 min) to inactivate the enzymes and subsequently stored at -80 °C until further analyses.

2.3 Batch maceration treatment of pectin at 50 °C

A pectin model solution, prepared as described for UAEM treatment, was incubated with 100 ppm of the enzyme preparation at 50 °C for one hour (**Figure 2-2**) according to the supplier's recommendation regarding application for fruit juice production, in the following referred to as batch maceration (benchmark). An aliquot of the solution without enzyme preparation was treated in the same way and was used as the control. Inactivation of the enzyme was carried out as described above. The experiments were repeated three times.

2.4 Characterization of the pectin model solution

2.4.1 Determination of the viscosity

The viscosity of pectin model solutions was determined with a rotational viscometer (V-Pad, Fungilab, New York City, NY) equipped with an LCP spindle. The analyses were run at 200 rpm for 1 min at 23 °C.

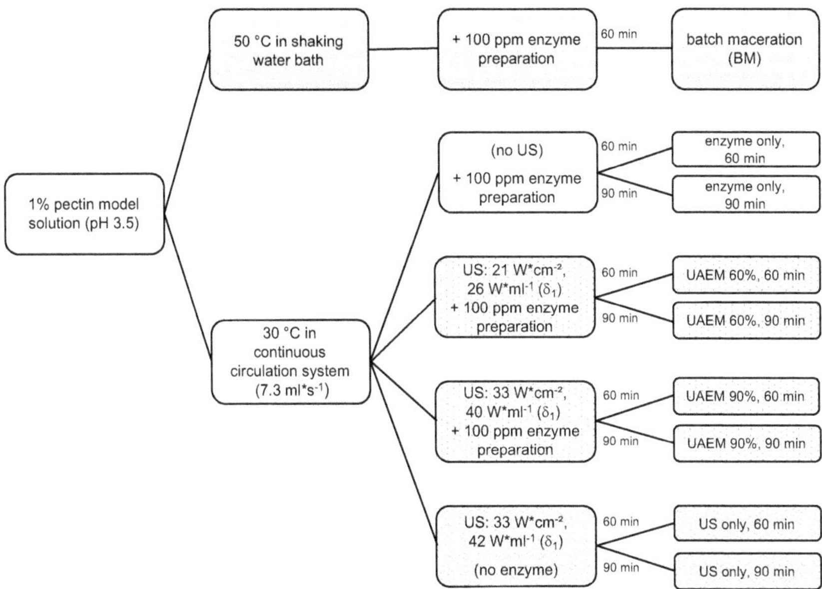

Figure 2-2: Experimental plan for treatment of pectin model solution (1% (w/v) in citrate buffer pH 3.5) by batch maceration at 50 °C or continuous circulation system at 30 °C (7.3 ml*sec^{-1}). The latter was carried out by enzyme, US, or combined (UAEM) application. Enzyme dosage was Rohapect®Classic 100 ppm. US power intensity (ml*cm^{-2}) and power density δ_1 (W*ml^{-1}) according to flow cell volume (9 ml) are given.

2.4.2 Determination of the molecular weight (MW) distribution by high-performance size exclusion chromatography (HPSEC)

The distribution of MW was analyzed by HPSEC with refractive index (RI) detection as described previously [15]. Pullulan standards (6.1-708 kDa, ReadyCal-Kit Pullulan, PSS-Polymer Standards, Mainz, Germany) were used to calculate MWs. The chromatograms were divided into three representative fractions and their relative peak areas were calculated: Fraction 1 (117-708 kDa), fraction 2 (30-117 kDa) and fraction 3 (6-30 kDa).

2.4.3 Determination of the degree of methylation (DM) and the degree of acetylation (DA)

DM and DA were determined by quantification of released methanol and acetic acid, respectively, by headspace solid-phase dynamic extraction gas chromatography (HS-

SPDE-GC) with flame ionization detection (FID) after saponification as described previously [15]. DM and DA were calculated as mol methyl/acetyl groups per 100 mol GalAc [28].

2.4.4 Quantification of galacturonic acid (GalAc), L-rhamnose and L-fucose in the low molecular weight fraction (<30 kDa)

Three specific monomers, GalAc, L-rhamnose, and L-fucose, were quantified after acid hydrolysis using the respective enzyme-kit from Megazyme (Wicklow, Ireland) [15]. Prior to acid treatment, samples were separated at room temperature by their MW using Vivaspin 15 R (MWCO 30 kDa, Sartorius, Goettingen, Germany) with a Heraeus Megafuge 40R Centrifuge (Thermo Fisher Scientific, Braunschweig, Germany) at 6000*g*. Monomers were analyzed in the low MW fraction and were compared to the respective amount in an untreated sample representing the total amount of the monomers in the native pectin.

2.5 Determination of enzyme activities

The assays for the determination of the specific enzyme activities (nkat*ml^{-1}) were adjusted to optimal conditions described below based on Li *et al.* [29]. Incubation time, substrate concentration, enzyme dilution, and the substrate were evaluated to obtain consistent results.

Additionally, these assays were used to examine the effective enzyme activities during batch maceration and treatments in the continuous circulation system under the respective conditions (**Figure. 2-2**). Incubation time (60 and 90 min), substrate concentration (1% w/v), and enzyme dilution (100 ppm) were based on the experimental approach and were considered in the corresponding calculations. The final marker concentration, determined in each assay, was analyzed after the whole process of 60 or 90 min to compare the total output of each process.

All experiments including the controls without enzyme addition were carried out in triplicate.

2.5.1 *Characterization of polygalacturonase (PG) activity*

As the marker for PG activity, the amount of reducing sugar released from pectin was determined. Reducing sugars were quantified by the 3,5-dinitrosalicylic acid method [30] with slight modifications. For the determination of the specific enzyme activity, apple pectin (low methylated) was used. The pectin suspension (300 µl; 0.5%, w/v) dissolved in sodium citrate buffer (0.1 M, pH 3.5) and 30 µl of aqueous enzyme solution (30%, v/v) were incubated for 30 min at 45 °C in a shaking water bath. The reaction was terminated by the addition of 1000 µl 3,5-dinitrosalicylic acid reagent. After boiling for 5 min, cooling on ice and centrifugation for 5 min at 9,600*g* (Heraeus Pico 17, Thermo Fisher Scientific, Braunschweig, Germany), the absorption was measured at 540 nm. PG activity was calculated as GalAc equivalents.

2.5.2 *Characterization of pectin lyase (PL) activity*

The marker of PL activity was the amount of formylpyruvate formed in alkaline media after enzymatic β-elimination from pectin. The aldehyde was quantified by the thiobarbituric acid method [31] with slight modifications. For the determination of the specific pectin lyase activity, sugar beet pectin was used. The pectin suspension (250 µl; 0.05%, w/v) dissolved in sodium citrate buffer (0.1 M, pH 3.5) and 30 µl of aqueous enzyme solution (30%, v/v) were incubated for 30 min at 45 °C in a shaking water bath. For the formation of formylpyruvate, 300 µl of sodium hydroxide solution (1 M) was added and incubated for 5 min at 80 °C and cooled on ice (1 min). Subsequently, 360 µl of hydrochloric acid (1 M) and 300 µl of thiobarbituric acid (0.04 M) were added and the solution was incubated 30 min at 80 °C to obtain a pink fluorescent dye. After cooling on ice and centrifugation for 5 min at 9,600*g* (Heraeus Pico 17, Thermo Fisher Scientific, Braunschweig, Germany), the absorption was measured at 550 nm. Different concentrations of malondialdehyde, which is of a comparable structure and shows similar absorption characteristics to formylpyruvate, were used for external calibration [32,33].

2.5.3 *Characterization of pectin methylesterase (PME) activity*

The released methanol was the analyzed marker for PME activity. The concentration of methanol was quantified by headspace solid-phase dynamic extraction gas chromatography (HS SPDE GC) with flame ionization detection (FID) as described previously [15]. For the enzymatic saponification, 1600 µl apple pectin suspension (0.5%, w/v, high methylated) and 30 µl aqueous enzyme solution (30%, v/v) were incubated for 30 min at 45 °C. Enzymatic reactions were stopped by heat treatment (100 °C, 3 min) and samples were cooled on ice. An aliquot of 1300 µl was transferred to a 10 ml GC vial and mixed with 100 µL *n*-propanol (0.1% w/v) used as an internal standard. External calibration was based on methanol standard solutions (0.125-5% w/v) and was corrected by *n*-propanol as the internal standard (0.1% w/v).

2.6 Statistical analysis

To determine significant differences, an ANOVA using Bonferroni post-hoc test was performed by XLSTAT software version 2014.4.06 (Addinsoft, Paris, France). The level of significance was defined as $p \leq 0.05$.

3 Results and Discussion

3.1 Specific enzyme activities of the preparation

The three specific activities of the commercial enzyme preparation that are PG, PL, and PME were determined (**Table 2-1**) under adjusted conditions based on Li *et al.* [29], who proposed a reproducible assay independent of dilution factors. Based on preliminary tests, the optimal incubation time and the optimal temperature was 30 min at 45 °C for all assays using a specific substrate concentration of 5 $g*L^{-1}$ and enzyme preparation amounts of 3% (v/v) for PG assay, 3.6% (v/v) for PL, and 0.56% (v/v) for PME, respectively. The results were in range with other commercial pectinolytic enzyme preparations for fruit juice maceration, where PG and PME activities are usually the most dominant. [9] Pectinases are classified according to the catalyzed reaction during pectin breakdown and grouped into three main categories: lyases, hydrolases, and esterases. PL (E.C. 4.2.2.10) catalyzes the cleavage of the 1,4-glycosidic bond of esterified GalAc by trans elimination leaving unsaturated residues. PG (E.C. 3.2.1.15) catalyzes the hydrolytic cleavage of 1,4-glycosidic bonds between non-esterified GalAc. PME (E.C. 3.1.1.11) catalyzes the de-esterification of the methoxy groups of pectin. [34] All three pectinolytic activities mainly act on the HG subunit consisting of polygalacturonic acid and are important for cell wall degradation in juice maceration due to the high pectin content of fruits and berries.

Table 2-1. Main specific pectinolytic activities of enzyme preparation Rohapect®Classic.

	PG[a]	PME[b]	PL[c]
specific activity[d] (nkat*ml^{-1})	2983 ± 116	2714 ± 167	3.3 ± 0.2

[a] Total-PG: nmol reducing ends (GalAc)/second
[b] nmol methanol/second
[c] nmol formylpyruvat (malondialdehyde)/second
[d] determined at pH 3.5 and 45 °C

3.2 Pectin degradation during batch maceration (50 °C)

A batch maceration using the commercial pectinolytic enzyme preparation was conducted under maceration conditions commonly used in red berry juice production (1 h, 50 °C, pH 3.5). Since most of the above-mentioned positive effects of polysaccharide degradation (viscosity reduction, interactions with anthocyanins) have been shown for such conditions, the characterization of the arising polysaccharides and oligosaccharides of this batch maceration provided the benchmark for the subsequent application of UAEM. Comparing the UAEM-induced effects with the benchmark, significant changes compared to the standard conditions can be revealed. The used pectin contained a medium DM of approx. 50% (**Table 2-2**) providing esterified and non-esterified GalAc binding sides for the different specific pectinolytic activities.

The applied enzyme preparation degraded pectin extensively by specific enzymatic cleavage clearly demonstrated by the considerable decrease in viscosity of 70%, indicating a substantial reduction of the average MW of pectin (**Table 2-2**). A detailed profile of the resulting MW distribution after batch maceration of the arising oligosaccharides and polysaccharides was determined by HPSEC (**Figure 2-3a**), illustrating a high polydispersity that implies a high diversity of pectin residues. Separated into three MW fractions, the profile predominantly comprises low MW (6-30 kDa) and medium MW (30-117 kDa), while the high MW (117-708 kDa) fraction probably consists of remaining non-degraded pectin (**Figure 2-4**).

Table 2-2. Characterization of pectin model solution after individual treatments; samples treated in continuous circulation system at 30 °C were compared to batch maceration at 50 °C and untreated native pectin. Different letters indicate significant differences in each column due to treatment.

	sample	time (min)	power density δ_1 ($W*ml^{-1}$)	power density δ_2 ($W*ml^{-1}$)	viscosity ($mPa*s^{-1}$)	DM (%)	DA (%)	low MW fraction (<30 kDa)		
								GalAc [a] ($mg*g^{-1}$)	rhamnose [a] ($mg*g^{-1}$)	fucose [a] ($mg*g^{-1}$)
	native	-	-	-	7.42 ± 0.27^{a}	47.92 ± 2.92^{a}	17.12 ± 1.23^{a}	11.25 ± 3.42^{f}	0.58 ± 0.13^{d}	0.27 ± 0.09^{b}
	batch maceration	60	-	-	2.26 ± 0.01^{d}	$36.88 \pm 3.37^{b,c}$	13.88 ± 2.40^{b}	$183.97 \pm 18.21^{a,b,c,d}$	$0.93 \pm 0.19^{b,c}$	0.79 ± 0.22^{a}
continuous circulation system	enzyme only	60	-	-	2.38 ± 0.05^{d}	39.88 ± 3.71^{b}	16.39 ± 1.14^{a}	$182.14 \pm 5.55^{b,c,d}$	$0.90 \pm 0.08^{b,c}$	0.75 ± 0.13^{a}
		90	-	-	2.34 ± 0.03^{d}	$38.04 \pm 2.17^{b,c}$	17.06 ± 0.94^{a}	$191.18 \pm 6.82^{a,b,c}$	$0.92 \pm 0.06^{b,c}$	0.91 ± 0.05^{a}
	UAEM 60%	60	21.5	26.5	2.37 ± 0.03^{d}	40.11 ± 2.05^{b}	17.72 ± 1.22^{a}	$175.80 \pm 14.92^{c,d}$	$1.23 \pm 0.12^{b,c}$	0.80 ± 0.13^{a}
		90	21.2	26.1	2.31 ± 0.05^{d}	$35.95 \pm 2.98^{b,c}$	17.96 ± 1.36^{a}	205.15 ± 9.80^{a}	$1.37 \pm 0.12^{a,b}$	0.99 ± 0.27^{a}
	UAEM 90%	60	32.8	40.4	2.35 ± 0.02^{d}	$38.10 \pm 1.9^{b,c}$	17.75 ± 0.92^{a}	163.77 ± 24.70^{d}	$1.01 \pm 0.35^{b,c}$	0.74 ± 0.09^{b}
		90	33.0	40.6	2.29 ± 0.02^{d}	34.46 ± 2.24^{c}	17.03 ± 1.37^{a}	$203.08 \pm 18.27^{a,b}$	1.46 ± 0.33^{a}	0.86 ± 0.07^{a}
	US only	60	33.9	41.7	3.28 ± 0.10^{b}	47.98 ± 2.37^{a}	$15.39 \pm 1.78^{a,b}$	$29.27 \pm 5.35^{e,f}$	$0.74 \pm 0.06^{c,d}$	0.30 ± 0.36^{b}
		90	34.0	41.9	2.76 ± 0.06^{c}	47.03 ± 1.44^{a}	17.36 ± 0.96^{a}	22.73 ± 3.89^{e}	$0.83 \pm 0.12^{b,c}$	0.38 ± 0.04^{b}

[a] monomer sugar content is calculated as $mg*g^{-1}$ pectin, power density δ_1 is calculated with flow cell volume (9 ml), power density δ_2 is calculated with sample volume in continuous circulation system (600 ml)

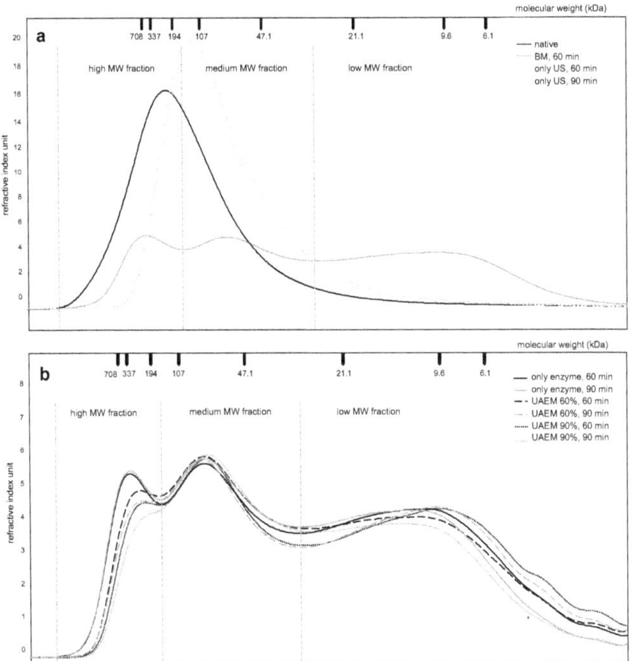

Figure 2-3. HPSEC elution pattern of (a) native pectin, continuously US treated (7.3 ml*sec^{-1}, 30 °C, A 90%, 300 W, 33 W*cm^{-2}, for 60 and 90 min) pectin, and pectin after batch maceration (BM, 50 °C, 100 ppm, 60 min) and (b) pectin after treatment in continuous circulation system (7.3 ml*sec^{-1}, 30 °C) by enzyme alone (100 ppm for 60 and 90 min), after UAEM 60% (100 ppm, A 60%, 190 W, 21 W*cm^{-2}, for 60 and 90 min), and after UAEM 90% (100 ppm, A 90%, 300 W, 33 W*cm^{-2}, for 60 and 90 min). Column was calibrated with pullulan standards to estimate the peak MW distribution. A: amplitude.

The medium MW fraction includes resistant fragments of pectin's hairy region, forming highly branched oligomers of RG I which have previously been structurally characterized by anion-exchange chromatography and SEC analysis [35]. Results are generally consistent compared to these authors, who degraded sugar beet pectin by PG and PME for 24 h. According to Ralet *et al.* [35], it can be suggested that a prolonged maceration time would probably reveal a greater decrease in high MW residues and a greater increase in low MW oligomers due to ongoing degradation of HG by pectinolytic enzymes. However, a certain amount of resistant fragments of medium MW remains after batch maceration since activities of common enzyme preparations cannot degrade the complex subunit of pectin's hairy region. [35]

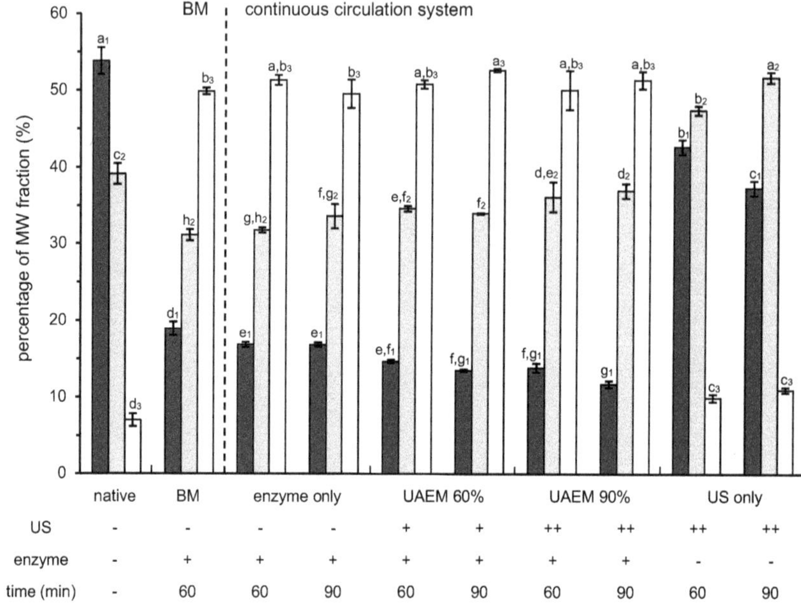

Figure 2-4. Percentage of MW (kDa) distribution yielded after treatments separated into high MW (117-708 kDa, dark grey bar), medium MW (30-117 kDa, grey bar) and low MW (6-30 kDa, white bar) fractions. Batch maceration (BM) was carried out at 50 °C. US only, enzyme only and UAEM were carried out in a continuous circulation system at 30 °C. Different letters indicate significant differences due to treatment. US ++: A 90%, 300 W, 33 W*cm^{-2} US +: A 60%, 190 W, 21 W*cm^{-2}, enzyme +: 100 ppm, enzyme -: no enzyme

Based on our previously published findings, oligomers of low MW form soluble complexes with polyphenols like anthocyanins leading to a stabilization of these value adding compounds. It was assumed that HG-derived residues stabilize anthocyanins rich in hydroxy groups, while RG-derived residues probably play a role in the complex formation of anthocyanins rich in methoxy groups. [15] Therefore, pectin model solutions and especially small oligomers in the low MW fraction were further characterized to examine degradation of pectin subunits into a set of corresponding oligomers by the quantification of GalAc, rhamnose, and fucose after MW separation by ultra-centrifugation (MWCO 30 kDa) in case of low MW fraction and acid hydrolysis (**Table 2-2**). GalAc is the predominant monosaccharide moiety in pectin and its content can be correlated with HG-derived residues. Rhamnose and fucose, due to their distinct occurrence in pectin substructures, can be related to the amount of RG I-derived and RG II-derived residues, respectively [36].

The absolute amount of these monomers in samples regarding all MW fraction was quantified without MW separation; the native samples contained 306.5 $\pm$ 18.1 mg*g^{-1} pectin of GalAc, 3.9 $\pm$ 1.3 mg*g^{-1} pectin of rhamnose and 2.3 $\pm$ 1.1 mg*g^{-1} pectin of fucose. Since the medium MW polymers did not form soluble complexes with anthocyanins in our previous study [15], they were not further characterized in the present study. Other sugar beet pectin components such as monomeric sugars (e.g., arabinose, galactose, glucose, and xylose), proteinaceous matter, ferulic acid, or metal ions were not further examined [37–39].The low MW fraction composition (<30 kDa, **Table 2-2**) show that enzyme activities mainly released GalAc containing residues, indicating a substantial degradation of HG chains to small oligomers. More than half (60.0%) of the absolute amount of GalAc was quantified in the low MW fraction (<30 kDa) after batch maceration, whereas an initial amount of 3.7% was found in the low MW fraction in the native pectin solution. These findings confirm the fact that PG and PL act as main activities by a high extent of a HG chain degradation into large amounts of polygalacturonic acid-residues [25,40]. The amounts of the minor monosaccharides rhamnose and fucose significantly increased in the low MW fraction, indicating a considerable release of RG I- and RG II-derived small oligomers by enzymatic side-activities like rhamnogalacturonase and *endo*-polygalacturonase (**Table 2-2**) [12]. Referred to the absolute amount in native pectin, 23.8% of rhamnose and 34.1% of fucose were detected in the low MW fraction (>30 kDa) after batch maceration, taking into account that the low MW fraction of native pectin solution already contained 15.0% of rhamnose and 11.7% of fucose. The findings demonstrate that pectin's subunits of RG I and RG II are difficult to degrade or resistant by enzymatic cleavage of the applied preparation. Thus, lower residue amounts of these subunits are degraded to low MW oligomers compared to HG-derived residues. The characterization of the arising oligomers under common batch maceration conditions is in agreement with previously reported findings. The enzymatic degradation has been shown to release polymers and oligomers including polyGalAc residues, alternating rhamnose-GalAc residues from RG I, RG II dimers, whereas complex RG I-derived residues that are resistant to the pectinase activities of common enzyme preparations remain as medium and high MW polymers [10,15,41].

The enzyme preparation contained a comparably high PME activity (**Table 2-1**) that reduced the DM of pectin by 23% during batch maceration. In our previous study, DM could be reduced enzymatically by >50% [15]. This difference can be explained by the different enzyme preparations applied varying in PME activity. Also, PME activity might be limited by substrate concentration due to the lower DM of the pectin used in this study. A high DM of pectin is known to increase gelling properties of pectin in an acidic environment by intermolecular hydrogen bridges between the partly undissociated carboxy groups and hydrophobic interactions of methylated carboxy groups of the galacturonic acids. [42] High viscosities are impeding juice production and need to be decreased. PME reduces pectin esterification, preparing binding sites for PG, which needs non-esterified galacturonic acid chains as substrate. The cleavage by PG consequently reduces viscosity by reducing the MW of polygalacturonic acid. Thus, a sufficient PME activity is necessary for an efficient pectin degradation by PG activity, especially of high DM pectin.

The DA was reduced by 18%, probably by side activity of an acetyl esterase (**Table 2-2**) that has been shown in other enzyme preparations for juice production before [15]. A reduction of DM and DA in pectin structure has been demonstrated to beneficially affect anthocyanin complexation. The reduction provides more binding sites (free galacturonic acid residues) and by lower steric hindrances to form hydrogen bonds or ionic interaction in acidic environments, due to dissociation of carboxylic acid groups (pK_a 3.5 [43]). Therefore, the characterization of DM and DA of pectin and pectin-derived residues provides important information on their ability to complex anthocyanins and other matrix compounds. [15]

3.3 Pectin degradation in a continuous circulation system (30 °C)

To obtain an enhanced pectin degradation into the described set of small oligomers by milder maceration conditions, a continuous circulation flow system with an implemented US probe in a flow cell was applied. Continuous systems have been demonstrated to be most convenient for juice production at an industrial scale. Applying US concurrently increases the temperature of the treated media mainly caused by the temperature of local hot-spots (5000 K) of collapsing cavitation bubbles [44]. By the use of a continuous circulation flow system with an implemented US probe in a cooled flow cell the temperature in the whole system could be controlled accurately at 30 °C ± 2 °C.

The effectiveness of combined enzymatic and US-induced pectin degradation as a milder maceration treatment was compared to the benchmark results of the batch maceration at 50 °C. The application of enzymes and US alone was investigated individually to differentiate between distinct and synergistic effects.

3.3.1 Exclusive enzyme treatment

Pectin degradation by the enzyme preparation in the continuous circulation system at reduced temperature (30 °C) was almost identical to the benchmark degradation during batch maceration at 50 °C for 60 min (**Figure 2-3**). All parameters but DA (**Table 2-2**) were equal to those of pectin degraded in batch maceration. The reduction of DA after treatment at 50 °C was more pronounced which can be attributed to a higher temperature optimum of the responsible acetyl esterase [45]. The observed results could be attributed to the fact that the main pectinolytic enzyme activities of the commercial preparation are very effective in pectin degradation even at the lower temperature of 30 °C.

3.3.2 Exclusive ultrasound treatment

In general, low-intensity US-induced effects non-specifically degrade polysaccharides by mechanical effects, in contrast to the enzymatic cleavage occurring specifically at distinct points of action in the substrate. This leads to a fundamentally different MW distribution (**Figure 2-3a**) of degraded pectin oligosaccharides and polysaccharides after the exclusive US treatment compared to the enzymatic benchmark hydrolysis. US treatment decreased the average MW of the native pectin with a molecular weight of the peak (Mp) of approx. 300 kDa into smaller polymers with a medium Mp of approx. 150 kDa. The mechanical forces of the applied low-intensity US creates cavitational collapses that induces shear forces and micro streams, breaking primarily chemical bonds of polymers by mechanical torque [25,46].

The narrow MW distribution of US modified pectin indicates a low polydispersity index, meaning that the homogeneity of MW distribution increased [46]. One attribute of US-induced polysaccharide degradation is the high fragmentation rate of large molecules to a narrow and more uniform MW distribution. This phenomenon is explained by a non-random breakage occurring approximately in the midpoint of the oscillating polymers induced by US pressure waves. The final length of the polymers depends on the intensity of the mechanical torque that is necessary to rupture the linkages. The degradation of high MW polymers occurs faster due to the less energy required for their breakage and, thus, the resulted pool of polymers presents a narrow MW distribution. [47] This phenomenon is also consistent with the following observations. The MW profile of degradation products mainly encompasses medium MW polymers (30-117 kDa; **Figure 2-4**); within prolonged treatment time (90 min) this fraction only slightly increased further (4.3%). The amount of low MW fraction of small oligomers (< 30 kDa) increased by only 1% over treatment time of 60 and 90 min. The results indicate that the applied energy in the continuous system was only sufficient for the production of medium MW polymers. The findings confirm the fact that the lower the MW of the polymers, the higher energy is necessary to require further breakage [48]. However, it has been claimed that an increase in US intensity only results in higher degradation effects up to a certain level. This is explained by hindrance of the energy transmission in the system as a consequence of an increased number of cavitation bubbles. A limit of up to 50 kDA for dextran and acidic polysaccharides [47], and 100 kDa for chitosan [49, 50] after prolonged treatment times was proposed.

The monosaccharide composition of small oligomers in the low MW fraction (<30 kDa, **Table 2-2**) revealed that only the amount of HG-derived and RG I-derived residues increased significantly, resulting in an increase of 5.9% GalAc and 6.5% rhamnose in the low MW fraction, referred to the initial content of native pectin solution. The amount of RG II-derived residues in the low MW fraction was not significantly changed compared to native pectin. These findings indicate that the applied low-intensity US broke the HG chain into only few small polygalacturonic acid residues and released RG I residues from the main chain. This result agrees with previously reported findings, where the authors explained the US-induced reduction MW of pectin was mainly due to the release of RG-I domains [36,51]. In accordance with literature, monosaccharides did not emerge during exclusive US treatments [48,51]. These findings confirm the fact that US-induced effects only reduce MW of polymers into smaller ones without the liberation of monomers and, consequently, without altering monosaccharide compositions of the produced juice.

The exclusive US treatment did not change DM and DA significantly (**Table 2-2**), probably because of the relatively low US intensity applied (max. 34 W*cm^{-2}) and the citric acid medium, which has been shown to inhibit the reduction of DM by US treatments [51]. In contrast, higher ultrasound intensity caused a significant reduction of DM and DA in previous studies [15,48]. High US intensities are apparently needed, if a reduction of DM or DA is necessary. In this study, higher US intensities were excluded to maintain mild maceration conditions at 30 °C.

Although US treated samples contained polymers of higher MW than that after those after enzymatic treatment, the decrease in viscosity was similar to enzymatic treatment by 62.8% (**Table 2-2**). Since viscosity is associated with high MW polymers, the considerable reduction of viscosity by US treatment has to be explained by the reduction of agglomerates and structural changes that lead to more linear polymers [15,48]. However, the exclusive US treatment did not produce a similar set of oligosaccharides and polysaccharides like the benchmark enzymatic degradation, which are necessary to enrich juice with soluble fiber and evoke anthocyanin stabilizing properties. Despite the comparable reduction of viscosity, an exclusive US treatment cannot replace the common enzymatic maceration. Additionally, polymers with medium MW created by US-treatment were shown to form insoluble anthocyanin-complexes in our former study, that would deprive valuable anthocyanin compounds in final juice [15].

3.3.3 Ultrasound-assisted enzymatic maceration (UAEM)

As indicated above, a combined application of US and enzymes has been demonstrated to enhance plant material extraction and facilitate polymer degradation in certain experimental conditions [1,6,19]. Accordingly, the present study was focused on the characterization of the arising polysaccharides and oligosaccharides during UAEM using a continuous circulation system and applying a common enzyme preparation, because certain small HG-derived and RG-derived oligomers have been shown to determine juice quality.

The UAEM application resulted in a quite similar MW distribution pattern (**Figure 2-3b**) compared to that observed after enzymatic degradation alone indicating also a diverse pool of oligosaccharides and polysaccharides. However, the MW distribution obtained after UAEM treatment was shifted to the lower MW fractions (**Figure 2-3b**), meaning that higher amounts of smaller and simpler molecules are formed compared to the benchmark trial (**Figure 2-3a**, **Table 2-2**). Compared to enzymatic degradation alone in the continuous system, high MW fraction was decreased significantly for UAEM 60%: -13% and -20% after 60 and 90 min, respectively, and for UAEM 90%: -18% and -31% after 60 and 90 min, respectively. The medium MW fraction was increased for UAEM 60%: +9% and +1% after 60 and 90 min, respectively, and for UAEM 90%: +14% and +10% after 60 min and 90 min, respectively, **Figure 2-4**. The medium MW fraction increased significantly more after shorter treatment of 60 min compared to the longer treatment of 90 min, due to the ongoing degradation of medium MW oligomers to smaller fragments. Although the MW distribution patterns after UAEM treatment qualitatively correlated with the enzymatic degradation patterns, degradation to polymers of medium MW and oligomers of smaller MW was enhanced with an increase in the applied US intensity. Thus, the higher US intensity (90%) revealed greater degradation effects than the lower (60%) in the UAEM system. It can be concluded that US affects enzymatic degradation, and that higher US intensities provoked more pronounced effects. The latter is explained by higher energy delivered into the system with increasing US intensity, thus increasing US-induced effects like cavitation and micro streaming [46,48].

Since the MW distribution of the formed oligosaccharides and polysaccharides is of great importance for stabilizing effects, it is noteworthy that US not only enhanced enzymatic degradation of pectin but also altered the MW distribution pattern, resulting in a profile of degradation products with structures different from those arising from enzymatic treatment alone. This effect can be explained by several mechanisms, like the US-induced alteration of the substrate, the US-induced alteration of the enzyme conformation and action, and the US-induced effects on the surrounding media. [24,36,52] This last effect is a more general effect on the overall reaction speed as it increases turbulences caused by pressure fluctuations that lead to an improved mass transfer accelerating substrate-enzyme exchange. Since the other two mechanisms are more likely to be specific for UAEM, they are discussed in the following.

3.3.4 Ultrasound-induced alteration of the substrate pectin

The findings of the exclusive US treatment demonstrated, that US degraded high MW polymers mainly into high and medium MW polymers. UAEM accordingly produced a significantly larger fraction of medium MW than enzyme treatment alone. The medium MW fraction after UAEM 90% increased more than after UAEM 60% like in the exclusive US treatment (***3.3.2***) indicating a prevalent US-induced production of medium MW polymers. However, it was reported that high cavitation phenomena deteriorate the efficiency of energy transmission into the solvent, reducing US-induced effects when the intensity exceeds a certain level. Zhang *et al.* [46] reported reduced US degradation effects on apple pectin at intensities >302 W*cm^{-2}. In the present experiments, the intensity was maintained sufficiently low (max. 34 W*cm^{-2}) to avoid such effects. As explained in section *3.3.2*, the shear forces of cavitation and micro-jets created by US change the structure of polymers and, thus, alters the substrates; it forms linear oligomers and breaks the complex, voluminous structure of pectin by concomitantly exposing more accessible domains that can be reached more easily by the enzymes. Consequently, enzymes degraded the US-modified polymers further into smaller molecules. [7,36] Thus, samples after UAEM treatment showed a reduced medium MW fraction compared to the medium MW fraction after US treatment alone, due to the additional enzymatic degradation, as can be seen from **Figure 2-4**. Concurrently, all UAEM treated samples presented highest amounts of small oligomers in the low MW fraction that consists of the highest amounts of GalAc, rhamnose and fucose monosaccharides (**Table 2-2**). After UAEM 90% (90 min) treatment, the content of the small oligomers (>30 kDa) were further elevated. The low MW fraction contained 66.3% of the absolute GalAc content, 37.5% of rhamnose and 37.0% of fucose, respectively. This distribution indicates that UAEM released more HG-derived, RG I-derived, and RG II-derived residues into the low MW fraction with increasing US intensity than the benchmark enzyme degradation at 50 °C. These findings are in agreement with Ma *et al.* [36], who pointed out that sonochemical-enzymatic effects degrade citrus pectin into smaller and simpler oligomers (batch trail, 4.5 W*ml^{-1}) with strongest synergistic effects at temperatures below 40 °C. The authors reported a higher decomposition of the HG subunits and an increase in contents of RG-I residues of degraded products by sonoenzymolysis reactions compared to enzymatic degradation of pectin. The pectin residues were shown to be less branched proven by atomic force microscopy [36]. In contrast, Muñoz-Almgro *et al.* [51] reported a decreased degradation of apple and citrus pectin by an US-assisted pectinase treatment compared to simple enzymatic degradation suggesting an inactivation of enzyme activities by ultrasound forces. Different experimental conditions like higher temperature (50 °C), higher US intensities (81.7 W*cm^{-2}), and different enzyme activities could explain these contrasting results in comparison with the present study.

The above reported mechanisms might suggest that US-induced effects simply prepare pectin for a subsequent enzymatic degradation. Our own preliminary tests and further studies on US pre-treatment before enzyme degradation led to inconsistent results. Lieu *et al.* [1] noticed enhanced cell wall degradation and juice yield by US-pretreatment of grape mash treatment, whereas Dalagnol *et al.* [24] and our preliminary results did not show any enhanced pectinase activity by US pre-treatment of apple and sugar beet pectin, respectively. Feng *et al.* [53] described an enhanced papain activity after the US pretreatment of sodium alginate due to favorable substrate-enzyme conformations. This discrepancy in the observed effects corroborates the conclusion that more than one mechanism is responsible, i.e. another mechanism apparently also plays a role for the observed synergistic effects, that is, the US-induced enzyme modification.

3.3.5 Ultrasound-induced effects on the enzyme

Mechanisms comparable to US-induced effects cleaving glycosidic bonds of the pectin chains also alter the conformation of enzymes during US treatment. The alterations of amino acid residues, the breakage of hydrogen bonding, and van der Waals interactions in polypeptide chains result in the modification of protein conformation and change the enzymes' catalytic activity. The tertiary and secondary structure of PG was studied in detail after US treatment (0.9 - 9 W*ml^{-1}, 30 °C, 40 min) by intrinsic fluorescence and circular dichroism spectra [7]. The authors reported an irreversible structural unfolding and a slight increase in β-sheet conformation that resulted in the formation of more active binding sites. The consequent changes of enzyme conformation thereby increased the activity. In contrast to these positive effects, negative effects of the alteration of proteins in terms of denaturation and inactivation have also been reported [51]. Muñoz-Almagro *et al.* [51] explained this by molecular unfolding, which was evaluated by intrinsic fluorescence due to tryptophan residues in pectinase. This contrasting results might be explained by harsher experimental conditions (81.7 W*cm^{-2}, 60 min, 50 °C) leading to enzyme denaturation.

Another theory attributes the improved enzyme activity to US induced hot-spots causing locally increased temperature, and, thereby shortly reaching enzyme optima [52]. This is substantiated by the observation that pectinase activity was enhanced by US mostly at temperatures below 40 °C which was indeed observed in preliminary experiments as well as in several studies [24,36]. At higher temperatures close to the optimum temperature of the enzymes (45-50 °C), US effects became negligible. A further increase in temperature (>60 °C) during US-treatment might even affect pectinases' conformation adversely and lower cavitational effects. The latter is explained by an increase of vapor pressure in the solution due to the increasing temperature leading to a decrease of viscosity and surface tension reducing collapse temperature. [19,36,51,54] The temperature in our experiments was thoroughly controlled by a cooling system and kept at 30 °C ± 2 °C to avoid such negative effects.

DM was only slightly reduced by UAEM (**Table 2-2**), which might be related to an impaired PME activity as discussed in detail in *3.4*. However, longer treatments (90 min) decrease DM values more (25% by UAEM 60% and 28% by UAEM 90%) than the enzymatic treatment of the benchmark trail (60 min, 23% by batch maceration). This means that PME is continuously active releasing more methoxy groups at prolonged reaction time. DA values of pectin were not affected by any UAEM treatment (**Table 2-2**), which demonstrates that US application has no synergistic effects on the acetyl esterase activity. The findings are in agreement with those previously reported, where the combined enzyme and US treatment of citrus pectin had no influence on the DA [36].

Regarding the exclusively enzymatic treatments in the benchmark trail, DA was only reduced by acetyl esterase at 50 °C. Therefore, it can be assumed that the hot-spot theory (momentarily reaching temperature optima) might be less important in case of acetyl esterase than conducive conformational modifications of the enzymes. While DA was significantly decreased solely by acetyl esterase at 50 °C, local hot-spots did not stimulate enzyme activity at 30 °C. The exact mechanisms of US-induced effects on the conformation of acetyl esterase still needs to be investigated.

Regarding the quality of degradation, the present results demonstrate that UAEM in a continuous flow system can be an effective method to produce a set of smaller polymers or oligomers with a greater diversity, which could be beneficial for fruit juice processing. The applied US treatment alone did not reveal a comparable degradation, but US holds synergistic effects by modification of both, substrate and enzyme, leading to a pronounced pectin degradation. Smaller oligomers that are not removed during juice filtration increase fiber content in juices and are assumed to possess higher anti-oxidant activities compared to bigger molecules. [48] Furthermore, low MW oligomers stabilize anthocyanin contents in soluble complexes during storage. A set of HG-derived and RG-derived oligomers has been shown to beneficially complex most anthocyanins that differ in their number of hydroxy groups and methoxy groups. Additionally, a decrease in DM beneficially affects these interactions due to an increased number of free binding sites on the galacturonic acid residues and less steric hindrances. [15]

3.4 Influence of ultrasound on effective enzyme activities during maceration

The observed results support the theory that low-intensity US treatment synergistically affects enzyme activities for enhanced pectin degradation, which has also been reported previously for pectinases in general, but only examined for PG activity in particular [2,24,25]. To evaluate these effects for each main pectinolytic activity, the effective enzyme activity was determined individually for PG, PL and PME in the continuous circulation system under the process conditions applied (**Figure 2-5**).

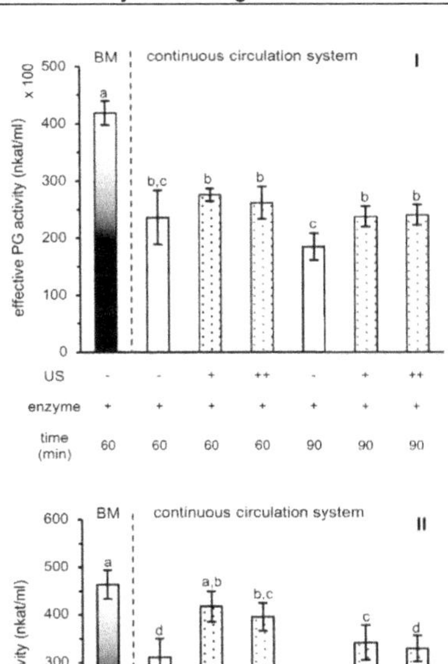

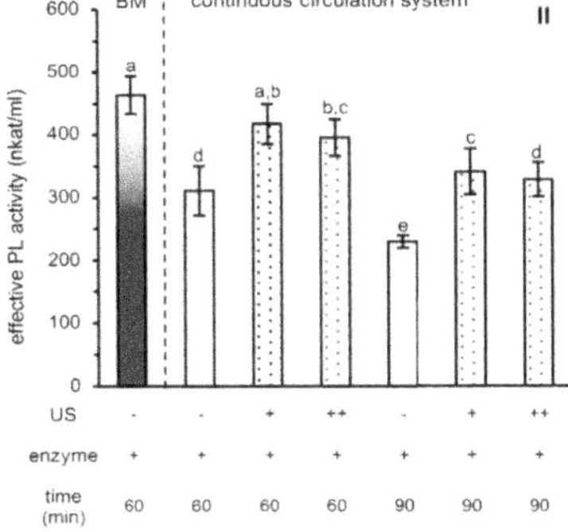

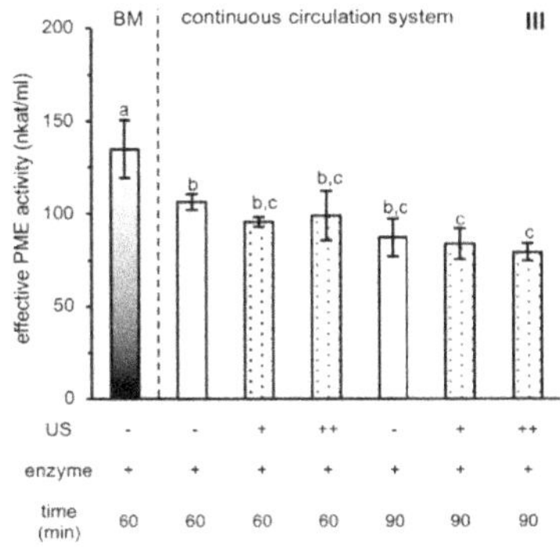

Figure 2-5. Effective enzyme activity (nkat/ml) of PG (I), PL (II) and PME (III) of the commercial preparation Rohapect®Classic (100 ppm) in pectin model solution (1%, pH 3.5) referred to the actual process conditions. Batch maceration (BM, grey bar) was carried out at 50 °C. US only (waved bar), enzyme only (white bar) and UAEM (dotted bar) were carried out in a continuous circulation system at 30 °C. Different letters indicate significant differences due to treatment. US ++: A 90%, 300 W, 33 W*cm^{-2} US +: A 60%, 190 W, 21 W*cm^{-2}, enzyme +: 100 ppm, enzyme -: no enzyme

The effective enzyme activities were generally higher during batch maceration at 50 °C (benchmark) compared to the continuous circulation system at 30 °C (**Figure 2-5**). These results are in agreement with the optimum temperature of the specific enzyme activity given by the manufacturer and previous studies [24,25]. The results obtained by the specific assays are somehow contradicting the results discussed for the UAEM effects, which clearly demonstrated a sufficient pectin degradation. The applied assays are used to characterize individual enzyme activities, but the degradation of pectin is a combination of all reactions. The observed differences between the lower distinct enzyme activities and the comparable pectin degradation can be explained by the marker compounds that are used for the assays, which do not necessarily reflect the effective degradation. However, the individual activities are observed to reveal the different synergistic effects of US on the three enzymes PG, PL and PME. The common assays only examine individual pectinolytic enzyme activity.

Regarding the individual enzyme effective activities, UAEM treatment significantly enhanced the PG and PL activities at 30 °C compared to enzymatic treatment alone within comparable treatment time (**Figure 2-5, I-II**). The results prove a synergistic effect of PL and US treatment that has not been published so far. Thus, the recently proposed synergistic effects of US on the "so called" pectinase activity [2,24,25], that has misleadingly only been proved for PG activity, also applies to PL activities. For the enzyme preparation used here, synergistic effects for PL were even higher than for PG. The effective enzyme activity of PL during UAEM 60% treatment was almost equal to the results of the benchmark trial at 50 °C (**Figure 2-5, II**). It has already been reported that similar US parameters have various effects on different enzymes. Yu *et al.* [55] reported the inactivation of α-amylase and papain under US irradiation, whereas the activity of pepsin was activated under the same experimental conditions. The authors attributed these deviant effects to the different secondary and tertiary structures of the enzymes. This could also apply to the individual conformations of PG and PL that are apparently influenced differently by US-induced effects. Both enzymes fold into a righthanded parallel β-helixes, but they differ in the number of β-sheets and the localization of disulfide bridges determining their conformation. In contrast to the three β-sheets in PL, the PG β-helix is comprised of four β-sheets. [56–58] Such β-sheets have been shown to be more prone to US irradiation, and might convert into α-helix, β-turn and random coil structures under higher US intensities (6 W^*ml^{-1}). [53,55,59] In contrast, at lower US intensities (2 W*ml-1, 15-30 min) an increase in β-sheets and a decreases of α-helix and random coil in soy protein isolate was observed under US treatment. [59] Also, Ma *et al.* [7] reported an increase of β-sheet and α-helix conformation after US treatment (US probe, 900 W, 9 W^*mL^{-1}, 40 min, 40 °C). The substrate binding site of PG is located on the exterior of the β-helix consisting of β-sheets indicating its essential conformation for catalytic cleavage. [7,60] The findings highlight the thin line between increasing and decreasing effects on enzyme activity by US treatment by either exposing or destruction of active sites, depending on the nature of enzymes conformation, US intensity, and other properties related with the ultrasound setup. Thus, within the US intensities applied in this study, PG and PL activity might be enhanced by conformational changes exposing of favorable β-sheet conformation. Although PG contains more β-sheets, PL activity was more enhanced suggesting other important conformational changes that have not been described so far. One reason might be that US-induced effects have been described to break disulfide bridges, decreasing intramolecular bindings of the enzymes. [53,59] Consequently, different amounts and localization of disulfide bridges, that are four in PG and two in PL, might affect conformation differently and change the enzymes' activity. [56,58] Subsequently, the more disulfide bridges are destroyed the more conformational changes might occur and thus adversely affect enzyme activity. A detailed investigation of conformational changes of the enzymes by different US intensity treatments is necessary to eventually prove this hypothesis.

The effective enzyme activities of PG and PL during the exclusive enzymatic treatment were lower after 90 min than after 60 min treatments (**Figure 2-5, I-II**), suggesting a partial inactivation of the enzymes over time. This loss of activity was compensated by UAEM treatments, which retained the effective enzyme activity over the longer periods of treatment for PG and PL. Similar effects were described by Ma *et al.* [7], who observed a prolonged lifetime and thermostability of PG activity by US treatment. For effective PL activity, UAEM 60% revealed slightly higher results than UAEM 90%, suggesting a greater synergistic effect at the lower US intensity (**Figure 2-5, II**). This fact could be attributed to a partial deactivation of PL by higher US intensities, since higher intensities may alter enzyme conformation detrimentally by more free radicals and stronger shear forces resulting in an denaturation and inactivation [36,61].

In contrast to PL and PG, the effective activity of PME was not enhanced by UAEM treatment. It was even slightly reduced by UAEM 90% application over 90 min (**Figure 2-5, III**). These findings suggest a possible specific sensitivity of PME activity toward US treatment, while other pectinase activities are improved. Tiwari *et al.* [62] reported the inactivation by 62% of PME even by even low-intensity US treatments (1.05 $W*mL^{-1}$) in orange juice. Fungal-PME folds into distinctive parallel β-helix structures containing even more β-sheets than the other two pectinases (PG and PL) encompassing several active-site residues. Additionally, it contains a cysteine ladder facing internally into the core of the β-helix. [63] Besides the above discussed vulnerability of β-sheets toward US irradiation, which can be adversely altered, US-induced oxygen radicals are assumed to oxidize free sulfhydryl groups forming sulfinic and sulfonic acids, destroying susceptible functional groups of the enzyme. [59,64] However, the applied US intensity in this study rather decelerated PME activity than inactivated it, since DM values of pectin was still significantly reduced after UAEM treatments compared to the native pectin (**Table 2-2**).

Enzymatic degradation of pectin is depended on the ensemble acting of all applied enzymatic activities and US. The observed results indicate a US-sensitivity of PME, consequently DM is only slightly reduced after incubation. However, PG activity requires de-esterified galacturonic acid residues as substrates and therefore depends on PME activity, if only methylated pectin is available. Consequently, PME inactivation entails a reduced PG activity. In contrast to PG, PL needs esterified galacturonic acid residues as substrates, which are more available, if PME activity is reduced by US application. This effect could also explain the higher PL activity rates during UAEM. To examine the US-induced effects on individual pectinolytic activities, specific tailormade enzymes with only one activity should be examined. However, in juice production enzyme preparations mainly produced by *Aspergillus Niger* encompass several activities that has to be considered for the complex pectin degradation.

The observed results demonstrate that the different enzyme activities of PL, PG and PME are affected individually by similar US parameters and experimental conditions. This dependency might be explained by the differently susceptible enzyme conformations leading to activation, deceleration or inactivation. Positive and negatives effects are acting at the same time summing up to a net-effect that can be observed and quantified.

3.5 Determination of total process output

Assessed only by the effective enzyme activities in the continuous circulation system at 30 °C compared to the benchmark in the batch maceration at 50 °C, it might be assumed that pectin degradation was less effective by UAEM. However, by considering the total output of the whole process, characterized by the corresponding final marker concentration determined in each assay (reducing sugars for PG [30], methanol for PME [28], and formylpyruvate for PL [31]), it can be seen that the UAEM treatments at 30 °C

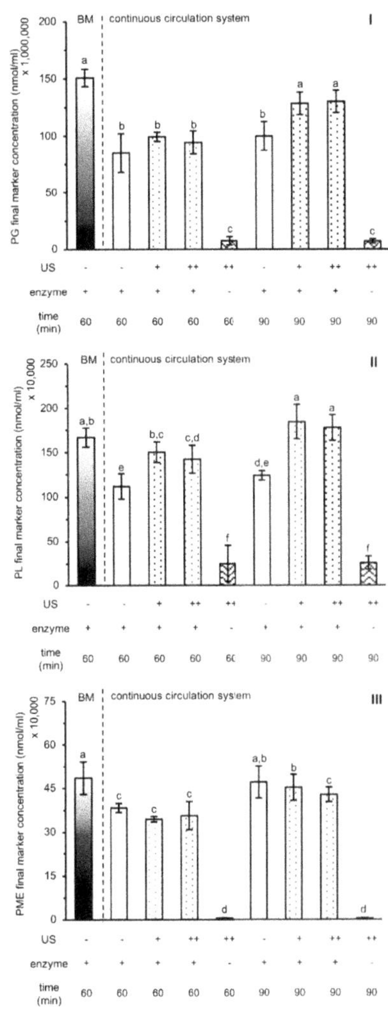

Figure 2-6. Final marker concentration (nmol/ml) of the specific enzyme assay of PG (I), PL (II) and PME (III) of the commercial preparation Rohapect®Classic (100 ppm) in pectin model solution (1%, pH 3.5) referred to the actual process conditions. Batch maceration (BM, grey bar) was carried out at 50 °C. US only (waved bar), enzyme only (white bar) and UAEM (dotted bar) were carried out in a continuous circulation system at 30 °C. Different letters indicate significant differences due to treatment. US ++: A 90%, 300 W, 33 W*cm^{-2} US +: A 60%, 190 W, 21 W*cm^{-2}, enzyme +: 100 ppm, enzyme -: no enzyme

lead to similar results as the benchmark at 50 °C (**Figure 2-6**). UAEM 90% treatments resulted in final marker concentrations for PG and PL equal to the benchmark, while quantities after UAEM 60% showed only a slightly lower level (**Figure 2-6, I-II**). Final marker concentration of PME was also higher after UAEM 90% treatments compared to UAEM 60%, but slightly lower compared to benchmark (**Figure 2-6, III**). This is explained by considering the final marker concentration over the whole process revealing comparable output concentrations at the end of each process, while the effective enzyme activities are expressed by rate of conversion per sec.

The results demonstrate that US affects enzyme activities at reduced temperature, rendering them at least as effective by revealing similar outputs in the applied process as at the given temperature optimum of the specific enzyme activity (50 °C, benchmark). Therefore, UAEM shows potential as an alternative processing technique for milder maceration conditions for juice production. A temperature reduction provides not only favorable conditions for heat-sensitive compounds, but also an energy-saving potential. Besides the similar total process output of enzymatic marker compounds compared to the benchmark, UAEM degraded pectin into a set of smaller oligomers derived from HG and RG subunits with reduced DM. The increased degradation of pectin may be promising for an increased cell wall degradation and juice yield in juice production. This primarily results in a higher extraction of plant compounds like polyphenols, but is also beneficial due to higher fiber contents in juice and the complexation of anthocyanins by UAEM-induced small oligomers to more stable complexes improving juice color and quality.

4 Conclusions

The applied enzyme preparation revealed effective pectin degradation in a continuous circulation system at reduced temperature (30 °C) regarding MW distribution, reduction of viscosity, and decrease in DM comparable to the batch maceration at 50 °C. UAEM in the continuous system resulted in an altered MW distribution by the formation of higher amounts of oligomers with medium MW (Mp approx. 150 kDa) and an enhanced release of HG-derived, RG I-derived, and RG II-derived smaller oligomers into the lower MW fraction (>30 kDa). The results suggest that the application of US affected enzymatic degradation by two mechanisms. On the one hand, it produced less complex oligomers mainly of medium MW with a high homogeneity of MW distribution, which could be subsequently degraded more easily by the enzymes. On the other hand, US affected enzyme's conformation resulting in synergistic effects that enhance the effective enzyme activities during process conditions for PG and PL by even prolonging their activity over time. The effective activity of PME was not enhanced by low-intensity treatment of UAEM, whereas UAEM 90% (33 $W{*}cm^{-2}$) slightly reduced the effective PME activity indicating a US sensitivity. The individual effects on enzyme activities might be explained by the different susceptibility of their conformation. Finally, the total process output of UAEM 90% treatments demonstrated for PG, PL and also nearly for PME that UAEM at reduced temperature was as efficient as the conventional batch maceration at 50 °C.

The potential of UAEM for degradation of pectin at reduced temperature offers several prospects for fruit juice production. Milder process conditions preserve heat-sensitive compounds, in particular vitamins and phenolic compounds including anthocyanins. Soluble small oligomers arising from sufficient pectin degradation add nutritional fiber to the juices. Modified polysaccharides like RG I residues have been described to act as health-promoting ingredients, showing antimicrobial and anti-cancer activity potential [21,22]. Low MW oligosaccharides provide an additional beneficial effect by complexing polyphenols like anthocyanins [15]; since these complexes prevent anthocyanins from oxidation and other degradation pathways, this may improve juice quality and nutritional value. [15,18] Ongoing research will need to verify our observed effects in actual juices, and might also be applied in other food matrices.

CRediT authorship contribution statement

Lena Rebecca Larsen: Conceptualization, Methodology, Investigation, Writing - original draft, Conceptualization, Methodology, Investigation, Writing - original draft. **Judith van der Weem:** Investigation, Formal analysis. **Rita Caspers-Weiffenbach:** Investigation, Formal analysis. **Andreas Schieber:** Resources, Funding acquisition. **Fabian Weber:** Conceptualization, Funding acquisition.

Declaration of Competing Interest

The authors declare no competing financial interest or personal relationships that could influence the reported work in this paper.

Acknowledgments

The authors gratefully thank AB Enzymes GmbH (Darmstadt, Germany) for providing the enzyme preparation and Herbstreith & Fox GmbH & Co. KG (Neuenbürg, Germany) for supplying pectin material.

Funding sources

The project is supported by funds of the Federal Ministry of Food and Agriculture (BMEL) based on a decision of the Parliament of the Federal Republic of Germany via the Federal Office for Agriculture and Food (BLE) under the innovation support program (281A102716).

References

[1] L.N. Lieu, V.V.M. Le, Application of ultrasound in grape mash treatment in juice processing, Ultrason. Sonochem. 17 (2010) 273–279. https://doi.org/10.1016/j.ultsonch.2009.05.002.

[2] W. Tchabo, Y. Ma, F.N. Engmann, H. Zhang, Ultrasound-assisted enzymatic extraction (UAEE) of phytochemical compounds from mulberry (*Morus nigra*) must and optimization study using response surface methodology, Ind. Crops Prod. 63 (2015) 214–225. https://doi.org/10.1016/j.indcrop.2014.09.053.

[3] S.R. Shirsath, S.H. Sonawane, P.R. Gogate, Intensification of extraction of natural products using ultrasonic irradiations – A review of current status, Chem. Eng. Process. Process Intensif. 53 (2012) 10–23. https://doi.org/10.1016/j.cep.2012.01.003.

[4] K.G. Zinoviadou, C.M. Galanakis, M. Brnčić, N. Grimi, N. Boussetta, M.J. Mota, J.A. Saraiva, A. Patras, B. Tiwari, F.J. Barba, Fruit juice sonication: Implications on food safety and physicochemical and nutritional properties, Food Res. Int. 77 (2015) 743–752. https://doi.org/10.1016/j.foodres.2015.05.032.

[5] M. Vinatoru, An overview of the ultrasonically assisted extraction of bioactive principles from herbs, Ultrason. Sonochem. 8 (2001) 303–313. https://doi.org/10.1016/S1350-4177(01)00071-2.

[6] N. Liao, J. Zhong, X. Ye, S. Lu, W. Wang, R. Zhang, J. Xu, S. Chen, D. Liu, Ultrasonic-assisted enzymatic extraction of polysaccharide from *Corbicula fluminea*: Characterization and antioxidant activity, LWT - Food Sci. Technol. 60 (2015) 1113–1121. https://doi.org/10.1016/j.lwt.2014.10.009.

[7] X. Ma, W. Wang, M. Zou, T. Ding, X. Ye, D. Liu, Properties and structures of commercial polygalacturonase with ultrasound treatment: Role of ultrasound in enzyme activation, RSC Adv. 5 (2015) 107591–107600. https://doi.org/10.1039/c5ra19425c.

[8] J. Buchert, J.M. Koponen, M. Suutarinen, A. Mustranta, M. Lille, R. Törrönen, K. Poutanen, Effect of enzyme-aided pressing on anthocyanin yield and profiles in bilberry and blackcurrant juices, J. Sci. Food Agric. 85 (2005) 2548–2556. https://doi.org/10.1002/jsfa.2284.

[9] I. Romero-Cascales, J.M. Ros-García, J.M. López-Roca, E. Gómez-Plaza, The effect of a commercial pectolytic enzyme on grape skin cell wall degradation and colour evolution during the maceration process, Food Chem. 130 (2012) 626–631. https://doi.org/10.1016/j.foodchem.2011.07.091.

[10] H. Hilz, E.J. Bakx, H.A. Schols, A. Voragen, Cell wall polysaccharides in black currants and bilberries - Characterisation in berries, juice, and press cake, Carbohydr. Polym. 59 (2005) 477–488. https://doi.org/10.1016/j.carbpol.2004.11.002.

[11] I. Oey, Effect of novel food processing on fruit and vegetable enzymes, in: A. Bayindirli (Ed.), Enzym. Fruit Veg. Process. - Chem. Eng. Appl., 1st ed., CRC Press Taylor and Francis Group, Boca Raton, 2010: pp. 245–312.

[12] J.A.E. Benen, A.G.J. Voragen, J. Visser, 67 Pectic enzymes, in: J.R. Whitaker, A.G.J. Voragen, D.W.S. Wong (Eds.), Handbook of food enzymology, Marcel Dekker New York, 2003: pp. 826-829

[13] A.G.J. Voragen, G.J. Coenen, R.P. Verhoef, H.A. Schols, Pectin, a versatile polysaccharide present in plant cell walls, Struct. Chem. 20 (2009) 263–275. https://doi.org/10.1007/s11224-009-9442-z.

[14] M. Holzwarth, S. Korhummel, T. Siekmann, R. Carle, D.R. Kammerer, Influence of different pectins, process and storage conditions on anthocyanin and colour retention in strawberry jams and spreads, LWT - Food Sci. Technol. 52 (2013) 131–138. https://doi.org/10.1016/j.lwt.2012.05.020.

[15] L.R. Larsen, J. Buerschaper, A. Schieber, F. Weber, Interactions of anthocyanins with pectin and pectin fragments in model solutions, J. Agric. Food Chem. 67 (2019) 9344–9353. https://doi.org/10.1021/acs.jafc.9b03108.

[16] F. Weber, L.R. Larsen, Influence of fruit juice processing on anthocyanin stability, Food Res. Int. 100 (2017) 354–365. https://doi.org/10.1016/j.foodres.2017.06.033.

[17] A. Padayachee, G. Netzel, M. Netzel, L. Day, D. Zabaras, D. Mikkelsen, M.J. Gidley, Binding of polyphenols to plant cell wall analogues - Part 1: Anthocyanins, Food Chem. 134 (2012) 155–161. https://doi.org/10.1016/j.foodchem.2012.02.082.

[18] M. Tomas, G. Rocchetti, S. Ghisoni, G. Giuberti, E. Capanoglu, L. Lucini, Effect of different soluble dietary fibres on the phenolic profile of blackberry puree subjected to *in vitro* gastrointestinal digestion and large intestine fermentation, Food Res. Int. 130 (2020) 108954. https://doi.org/10.1016/j.foodres.2019.108954.

[19] L.M.G. Dalagnol, L. Dal Magro, V.C.C. Silveira, E. Rodrigues, V. Manfroi, R.C. Rodrigues, Combination of ultrasound, enzymes and mechanical stirring: A new method to improve *Vitis vinifera* Cabernet Sauvignon must yield, quality and bioactive compounds, Food Bioprod. Process. 105 (2017) 197–204. https://doi.org/10.1016/j.fbp.2017.07.009.

[20] V.P.T. Nguyen, T.T. Le, V.V.M. Le, Application of combined ultrasound and cellulase preparation to guava (*Psidium guajava*) mash treatment in juice processing: Optimization of biocatalytic conditions by response surface methodology, Int. Food Res. J. 20 (2013) 377–381.

[21] F.W. Mohideen, K.M. Solval, J. Li, J. Zhang, A. Chouljenko, A. Chotiko, A.D. Prudente, J.D. Bankston, S. Sathivel, Effect of continuous ultra-sonication on microbial counts and physico-chemical properties of blueberry (*Vaccinium corymbosum*) juice, LWT - Food Sci. Technol. 60 (2015) 563–570. https://doi.org/10.1016/j.lwt.2014.07.047.

[22] E. Wong, F. Vaillant, A. Pérez, Osmosonication of blackberry juice: Impact on selected pathogens, spoilage microorganisms, and main quality parameters, J. Food Sci. 75 (2010). https://doi.org/10.1111/j.1750-3841.2010.01730.x.

[23] A.C. Soria, M. Villamiel, A. Montilla, Ultrasound Effects on Processes and Reactions Involving Carbohydrates, Ultrasound Food Process. Recent Adv. (2017) 437–463. https://doi.org/10.1002/9781118964156.ch17.

[24] L.M.G. Dalagnol, V.C.C. Silveira, H.B. da Silva, V. Manfroi, R.C. Rodrigues, Improvement of pectinase, xylanase and cellulase activities by ultrasound: Effects on enzymes and substrates, kinetics and thermodynamic parameters, Process Biochem. 61 (2017) 80–87. https://doi.org/10.1016/j.procbio.2017.06.029.

[25] X. Ma, L. Zhang, W. Wang, M. Zou, T. Ding, X. Ye, D. Liu, Synergistic effect and mechanisms of combining ultrasound and pectinase on pectin hydrolysis, Food Bioprocess Technol. 9 (2016) 1249–1257. https://doi.org/10.1007/s11947-016-1689-y.

[26] F. Chemat, N. Rombaut, A.G. Sicaire, A. Meullemiestre, A.S. Fabiano-Tixier, M. Abert-Vian, Ultrasound assisted extraction of food and natural products. Mechanisms, techniques, combinations, protocols and applications. A review, Ultrason. Sonochem. 34 (2017) 540–560. https://doi.org/10.1016/j.ultsonch.2016.06.035.

[27] C.M.G.C. Renard, M.J. Crépeau, J.F. Thibault, Structure of the repeating units in the rhamnogalacturonic backbone of apple, beet and citrus pectins, Carbohydr. Res. (1995). https://doi.org/10.1016/0008-6215(95)00140-O.

[28] S. Levigne, M. Thomas, M. Ralet, B. Quéméner, J. Thibault, Determination of the degrees of methylation and acetylation of pectins using a C18 column and internal standards, Food Hydrocoll. 16 (2002) 547–550.

[29] Q. Li, A.M. Coffman, L.K. Ju, Development of reproducible assays for polygalacturonase and pectinase, Enzyme Microb. Technol. 72 (2015) 42–48. https://doi.org/10.1016/j.enzmictec.2015.02.006.

[30] G.L. Miller, Use of dinitrosalicylic acid Reagent for determination of reducing sugar, Anal. Chem. 31 (1959) 426–428. https://doi.org/10.1021/ac60147a030.

[31] M. Nedjma, N. Hoffmann, A. Belarbi, Selective and sensitive detection of pectin lyase activity using a colorimetric test: Application to the screening of microorganisms possessing pectin lyase activity, Anal. Biochem. 291 (2001) 290–296. https://doi.org/10.1006/abio.2001.5032.

[32] K. Okamoto, C. Hatanaka, J. Ozawa, The thiobarbituric acid test of 4, 5-unsaturated digalacturonic acid, Berichte d. Ohara Instituts. 13 (1965) 7–12.

[33] A. Weissbach, J. Hurwitz, The formation of 2-Keto-3-deoxyheptonic acid in extracts of *Escherichia coli* B I. Identification, J. Biol. Chem. 234 (1959) 705–709.

[34] C. Grassin, Y. Coutel, 11 Enzymes in fruit and vegetable processing and juice extraction, in: R.J. Whitehurst, M. Van Oort (Eds.), Enzymes in food technologiy, 2nd edition, Wiley-Blackwell, 2010: p. 236.

[35] M.C. Ralet, J.C. Cabrera, E. Bonnin, B. Quéméner, P. Hellìn, J.F. Thibault, Mapping sugar beet pectin acetylation pattern, Phytochemistry. 66 (2005) 1832–1843. https://doi.org/10.1016/j.phytochem.2005.06.003.

[36] X. Ma, W. Wang, D. Wang, T. Ding, X. Ye, D. Liu, Degradation kinetics and structural characteristics of pectin under simultaneous sonochemical-enzymatic functions, Carbohydr. Polym. 154 (2016) 176–185. https://doi.org/10.1016/j.carbpol.2016.08.010.

[37] M. Buchweitz, A. Nagel, R. Carle, D.R. Kammerer, Characterisation of sugar beet pectin fractions providing enhanced stability of anthocyanin-based natural blue food colourants, Food Chem. 132 (2012) 1971–1979. https://doi.org/10.1016/j.foodchem.2011.12.034.

[38] P.A. Williams, C. Sayers, C. Viebke, C. Senan, J. Mazoyer, P. Boulenguer, Elucidation of the emulsification properties of sugar beet pectin, J. Agric. Food Chem. 53 (2005) 3592–3597. https://doi.org/10.1021/jf0404142.

[39] B.M. Yapo, C. Robert, I. Etienne, B. Wathelet, M. Paquot, Effect of extraction conditions on the yield, purity and surface properties of sugar beet pulp pectin extracts, Food Chem. 100 (2007) 1356–1364. https://doi.org/10.1016/j.foodchem.2005.12.012.

[40] C. Remoroza, S. Cord-Landwehr, A.G.M. Leijdekkers, B.M. Moerschbacher, H.A. Schols, H. Gruppen, Combined HILIC-ELSD/ESI-MSn enables the separation, identification and quantification of sugar beet pectin derived oligomers, Carbohydr. Polym. 90 (2012) 41–48. https://doi.org/10.1016/j.carbpol.2012.04.058.

[41] M.A. Ducasse, P. Williams, R.M. Canal-Llauberes, G. Mazerolles, V. Cheynier, T. Doco, Effect of macerating enzymes on the oligosaccharide profiles of merlot red wines, J. Agric. Food Chem. 59 (2011) 6558–6567. https://doi.org/10.1021/jf2003877.

[42] C.D. May, Industrial pectins: Sources, production and applications, Carbohydr. Polym. 12 (1990) 79–99. https://doi.org/10.1016/0144-8617(90)90105-2.

[43] R. Kohn, P. Kovác, Dissociation constants of D-galacturonic and D-glucuronic acid and their O-methyl derivatives, Chem. Zvesti. 32 (1978) 478–485.

[44] E.B. Flint, K.S. Suslick, The temperature of cavitation, Science 253.5026 (1991) 1397–1399. https://doi.org/10.1126/science.253.5026.1397.

[45] M.M. Kool, H.A. Schols, M. Wagenknecht, S.W.A. Hinz, B.M. Moerschbacher, H. Gruppen, Characterization of an acetyl esterase from *Myceliophthora thermophila C1* able to deacetylate xanthan, Carbohydr. Polym. 111 (2014) 222–229. https://doi.org/10.1016/j.carbpol.2014.04.064.

[46] L. Zhang, X. Ye, T. Ding, X. Sun, Y. Xu, D. Liu, Ultrasound effects on the degradation kinetics, structure and rheological properties of apple pectin, Ultrason. Sonochem. 20 (2013) 222–231. https://doi.org/10.1016/j.ultsonch.2012.07.021.

[47] S.C. Szu, G. Zon, R. Schneerson, J.B. Robbins, Ultrasonic irradiation of bacterial polysaccharides. Characterization of the depolymerized products and some applications of the process, Carbohydr. Res. 152 (1986) 7–20. https://doi.org/10.1016/S0008-6215(00)90283-0.

[48] J.K. Yan, Y.Y. Wang, H. Le Ma, Z. Bin Wang, Ultrasonic effects on the degradation kinetics, preliminary characterization and antioxidant activities of polysaccharides from *Phellinus linteus mycelia*, Ultrason. Sonochem. 29 (2016) 251–257. https://doi.org/10.1016/j.ultsonch.2015.10.005.

[49] R.H. Chen, J.R. Chang, J.S. Shyur, Effects of ultrasonic conditions and storage in acidic solutions on changes in molecular weight and polydispersity of treated chitosan, Carbohydr. Res. 299 (1997) 287–294. https://doi.org/10.1016/S0008-6215(97)00019-0.

[50] R.H. Chen, J.S. Chen, Changes of polydispersity and limiting molecular weight of ultrasonic treated chitosan, in: M.G. Peter, A. Domard, R.A.A. Muzzarelli (Eds.), Adv. Chitin Sci. Vol. 4, University of Potsdam, 2000: pp. 361–366.

[51] N. Muñoz-Almagro, A. Montilla, F.J. Moreno, M. Villamiel, Modification of citrus and apple pectin by power ultrasound: Effects of acid and enzymatic treatment, Ultrason. Sonochem. 38 (2017) 807–819. https://doi.org/10.1016/j.ultsonch.2016.11.039.

[52] M.M. Delgado-Povedano, M.D. Luque de Castro, A review on enzyme and ultrasound: A controversial but fruitful relationship, Anal. Chim. Acta. 889 (2015) 1–21. https://doi.org/10.1016/j.aca.2015.05.004.

[53] L. Feng, Y. Cao, D. Xu, S. You, F. Han, Influence of sodium alginate pretreated by ultrasound on papain properties: Activity, structure, conformation and molecular weight and distribution, Ultrason. Sonochem. 32 (2016) 224–230. https://doi.org/10.1016/j.ultsonch.2016.03.015.

[54] A.L. Prajapat, P.B. Subhedar, P.R. Gogate, Ultrasound assisted enzymatic depolymerization of aqueous guar gum solution, Ultrason. Sonochem. 29 (2016) 84–92. https://doi.org/10.1016/j.ultsonch.2015.09.009.

[55] Z.L. Yu, W.C. Zeng, W.H. Zhang, X.P. Liao, B. Shi, Effect of ultrasound on the activity and conformation of α-amylase, papain and pepsin, Ultrason. Sonochem. 21 (2014) 930–936. https://doi.org/10.1016/j.ultsonch.2013.11.002.

[56] S.K. Niture, Comparative biochemical and structural characterizations of fungal polygalacturonases, Biologia (Bratisl). 63 (2008) 1–19. https://doi.org/10.2478/s11756-008-0018-y.

[57] J. Vitali, B. Schick, H.C.M. Kester, J. Visser, F. Jurnak, The three-dimensional structure of *Aspergillus niger* pectin lyase b at 1.7-Å resolution, Plant Physiol. 116 (1998) 69–80. https://doi.org/10.1104/pp.116.1.69.

[58] O. Mayans, M. Scott, I. Connerton, T. Gravesen, J. Benen, J. Visser, R. Pickersgill, J. Jenkins, Two crystal structures of pectin lyase A from *Aspergillus* reveal a pH driven conformational change and striking divergence in the substrate-binding clefts of pectin and pectate lyases, Structure. 5.5 (1997) 677–689. https://doi.org/10.1016/S0969-2126(97)00222-0.

[59] H. Hu, J. Wu, E.C.Y. Li-Chan, L. Zhu, F. Zhang, X. Xu, G. Fan, L. Wang, X. Huang, S. Pan, Effects of ultrasound on structural and physical properties of soy protein isolate (SPI) dispersions, Food Hydrocoll. 30 (2013) 647–655. https://doi.org/10.1016/j.foodhyd.2012.08.001.

[60] G. van Pouderoyen, H.J. Snijder, J.A.E. Benen, B.W. Dijkstra, Structural insights into the processivity of endopolygalacturonase I from *Aspergillus niger*, FEBS Lett. 554 (2003) 462–466. https://doi.org/10.1016/S0014-5793(03)01221-3.

[61] X. Ma, J. Cai, D. Liu, Ultrasound for pectinase modification: an investigation into potential mechanisms, J. Sci. Food Agric. 100 (2020) 4636–4642. https://doi.org/10.1002/jsfa.10472.

[62] B.K. Tiwari, K. Muthukumarappan, C.P. O'Donnell, P.J. Cullen, Inactivation kinetics of pectin methylesterase and cloud retention in sonicated orange juice, Innov. Food Sci. Emerg. Technol. 10 (2009) 166–171. https://doi.org/10.1016/j.ifset.2008.11.006.

[63] L.M. Kent, T.S. Loo, L.D. Melton, D. Mercadante, M.A.K. Williams, G.B. Jameson, Structure and properties of a non-processive, salt-requiring, and acidophilic pectin methylesterase from *Aspergillus niger* provide insights into the key determinants of processivity control, J. Biol. Chem. 291 (2016) 1289–1306. https://doi.org/10.1074/jbc.M115.673152.

[64] İ. Gülseren, D. Güzey, B.D. Bruce, J. Weiss, Structural and functional changes in ultrasonicated bovine serum albumin solutions, Ultrason. Sonochem. 14 (2007) 173–183. https://doi.org/10.1016/j.ultsonch.2005.07.006.

Chapter 3

Interactions of anthocyanins with pectin and pectin fragments in model solutions

Anthocyanins determine the color and potential health-promoting properties of red fruit juices, but the juices contain remarkably less anthocyanins than the fruits, which is partly caused by the interactions of anthocyanins with the residues of cell wall polysaccharides like pectin. In this study, pectin was modified by ultrasound and enzyme treatments to residues of polysaccharides and oligosaccharides widely differing in their molecular weight. Modifications decreased viscosity and degrees of acetylation and methylation and released smooth and hairy region fragments. Native and modified pectin induced different effects on the concentrations of individual anthocyanins after short-term and long-term incubation caused by both hydrophobic and hydrophilic interactions. Results indicate that both pectin and anthocyanin structure influence these interactions. Linear polymers generated by ultrasound formed insoluble anthocyanin complexes, whereas oligosaccharides produced by enzymes formed soluble complexes with protective properties. The structure of the anthocyanin aglycone apparently influenced interactions more than the sugar moiety.

Keywords: anthocyanins, sugar beet pectin, pectinases, ultrasound treatment, size exclusion chromatography

Abbreviations: cya, cyanidin; DA, degree of acetylation; del, delphinidin; DM, degree of methylation; EMP, enzyme-modified pectin; GalAc, galacturonic acid; mal, malvidin; pel, perlagonidin; peo, peondin; pet, petunidin; SBP, sugar beet pectin; TAC, total anthocyanin content; UMP, ultrasound-modified pectin

This chapter has been published:

Larsen, L. R., Buerschaper, J., Schieber, A., & Weber, F. (**2019**) Interactions of anthocyanins with pectin and pectin fragments in model solutions. *Journal of Agricultural and Food Chemistry*, *67*(33), 9344-9353.

1 Introduction

Red berries and the derived juices are rich in anthocyanins that are primarily responsible for the appealing red to purple color and are also associated with numerous potential health benefits.[1,2] Red juices produced from these fruits contain significantly less anthocyanins compared to the raw material due to several processing steps that entail phenol degradation of these molecules.[3] The highest proportion of anthocyanins is lost during maceration and pressing. Anthocyanins are either insufficiently extracted or complexed by matrix compounds such as cell wall polysaccharides and, thus, a high quantity of anthocyanins remains in the press cake.[4–6]

Despite being very important for anthocyanin yield in the eventual juice, these interactions are not well-understood so far. Juice production requires efficient cell wall degradation, which is commonly achieved by the application of specific enzyme preparations containing various pectinases with further side activities. The advantages are an increased juice yield, lower juice viscosity, and increased extraction of bound anthocyanins. The advantages are an increased juice yield, lower juice viscosity, and increased extraction of bound anthocyanins.[7,8] Ultrasound technology provides another method to degrade plant cell walls by shear forces and cavitation.[9] While this technique is mainly used for pasteurization, it also shows a high potential for gentle extraction of anthocyanins.[10]

Plant cell walls consist of numerous polysaccharides and those of red berries contain a notable higher proportion of pectin.[11] Pectin contains two main structural elements: homogalacturonan (HG, ~60%) and rhamnogalacturonan I (RG I, ~20–35%). HG is composed of a linear chain of 1,4-linked galacturonic acid (GalAc, minimum 72–100 residues) that can be methylated at C-6 and acetylated at positions *O*-2 and *O*-3, expressed as the degree of methylation (DM) and acetylation (DA), respectively.[12] RG I has a backbone of alternating rhamnose and GalAc residues, while 20-80% of GalAc is attached to neutral sugar side chains like galactans, arabinans and arabinogalactans type I and II.[13] Rhamnogalacturonan II (RG II) is a minor component of pectin (0.5-8%) attached to HG. Its backbone consists of 8–10 GalAcs with four complex side chains consisting of 12 different monomers including rare sugars like fucose.[14]

Pectin polysaccharides have been demonstrated to interact with anthocyanins by weak bonds like hydrophobic forces and hydrogen bonds.[15] The latter are formed between hydroxy groups of anthocyanins and nonesterified GalAc in the pectin structure. The pH of juices reinforces these interactions, due to the pH-dependent equilibria of anthocyanins to flavylium cations and the negatively charged dissociated carboxylic acid groups of pectin.[16,17]

During fruit juice production, several pectin fragments are generated that may interact with extracted anthocyanins. Like the enzyme preparations used for juice production, ultrasound treatment degrades cell walls and produces several polymers with varying molecular weight (MW), sugar composition, or DM and DA. Here, treatment type, dosage or energy input, treatment time, and cell wall structure influence the resulting blend of numerous different polysaccharides and oligosaccharides.[18,19] Due to the modification of pectin polysaccharides by these different treatments, the interactions toward anthocyanins will be changed, which has not been investigated so far.

Previous studies have focused on the interaction of native pectins from various sources and selected anthocyanins[16,20,21], but generally lack information on the effects of juice processing on these interactions. The present work investigates similar interactions observed between modified pectin fragments, which are produced in the maceration step, and individual anthocyanins. These effects on a broad profile of anthocyanins were studied in a model solution to obtain a better understanding of the molecular drivers of anthocyanin−pectin interactions, which can lead to the mentioned anthocyanin losses during juice production. Sugar beet pectin resembles berry pectin rather than citrus or apple pectin regarding composition and characteristics.[5,20] Pectin was modified by two different approaches to generate distinct pectin fragments. Ultrasound modified pectin (UMP) and enzyme modified pectin (EMP) were incubated for 2 h or 2 weeks with two anthocyanin mixtures varying in their aglycone and glycoside compositions.

2 Materials and Methods

Chemicals, Reagents, and Standards. Ultrapure water was obtained from a PURELAB flex 2 water-purification system (ELGA LabWater, Paris, France). Acetonitrile (HPLC grade), ethanol (99.7%), and acetic acid were obtained from VWR (Mannheim, Germany).

Ethanol (HPLC grade) and citric acid monohydrate were purchased from Carl Roth GmbH & Co. KG (Karlsruhe, Germany). Methanol (HPLC grade) and sulphuric acid (95%) were from Th. Geyer (Renningen, Germany) and cyanidin-3-*O*-glucoside (>97.0%) was from Phytoplan (Heidelberg, Germany). Formic acid (99.9%) was obtained from Sigma-Aldrich (St. Louis, MO). The compounds m-phenylphenol, *n*-propanol, propionic acid, and sodium azide were purchased from Merk (Darmstadt, Germany), and sodium hydroxide from Honeywell (Morris Plains, NJ). Potassium sorbate (>99%), sodium tetraborate decahydrate, and D-(+)-GalAc monohydrate (99%) were obtained from Fluka (Munich, Germany), sodium nitrate (99%) from Acros Organics (Geel, Belgium), and trisodium citrate dihydrate and n-butanol (99%) were from Alpha Aesar (Ward Hill, MA).

Enzymes and Assay Kit. Enzyme preparation Klerzym150 and RohapectMA Plus were kindly provided by DSM Food Specialities B.V. (Heerlen, The Netherlands) and AB Enzymes GmbH (Darmstadt, Germany), respectively. L-Fucose assay kit, D-glucuronic acid/D-GalAc assay kit, and L-rhamnose assay kit were purchased from Megazyme (Wicklow, Ireland).

Preparation of Pectin Model Solution and Modified Pectin Residues. Sugar beet pectin (SBP) Betapec RU 301 was kindly provided by Herbstreith & Fox (Neuenbürg, Germany). SBP was dissolved in a 0.05 M sodium citrate buffer (pH 3.5) at a concentration of 0.75% (w/v) by stirring the suspension overnight at 60 °C. SBP was modified by an enzyme or ultrasound treatment to produce the pectin residues EMP and UMP, respectively. The two enzyme preparations are commonly applied for berry juice production and were used at a dosage of 100 ppm. Incubation was performed in sealed flasks in a shaking water bath (80 rpm, 40 °C) for 1, 2, and 4 h. Enzymes were inactivated at 100 °C for 3 min. The experiments were carried out in triplicate. Ultrasound treatments were performed with an ultrasound probe processor (UIP 1000hdT, 1000 W, 20 kHz, Hielscher, Teltow, Germany) equipped with sonotrode (9.0 cm^2) and booster horn (100% amplitude: 53 μm). To generate UMP, the ultrasound probe was immersed 2 cm below the liquid level in pectin solution (60 mL). Treatment was run at 60% amplitude (pulse duration 2 s) for 40 or 150 min and cooled on ice to keep the temperature below 40 °C. Specific energy input did not exceed 4.9 $W \cdot s^{-1} \cdot mL^{-1}$, and maximum energy density was 33 $W \cdot cm^{-2}$. The experiments were carried out in triplicate.

Pectin Characterization. The total uronic acid content was determined by the *m*-hydroxydiphenyl assay.[22]

DM and DA were determined as described in the literature with slight modifications.[23,24] Methanol and acetic acid were quantified by headspace solid-phase dynamic extraction gas chromatography (HS SPDE GC) with flame ionization detection (FID) after hydrolysis with 2 M NaOH. The SPDE equipment (Chromtech, Idstein, Germany) was installed in a CTC-Combi-PAL-Autosampler (Bender and Hobein, Zurich, Switzerland) to a GC FID system (Agilent Technologies model 6890). A SPDE needle (PDMS/AC/DVB coating, 50 mm × 0.8 mm, 0.53 mm) attached to a 2.5 mL gastight syringe (Hamilton, Darmstadt, Germany) pumping 50 cycles (100 $\mu L{\cdot}s^{-1}$) at a vial temperature of 55 °C was used for extraction. Splitless injection was performed at 250 °C onto an OPTIMAWAXplus column (30 m × 0.25 mm, 0.25 µL, Marcherey-Nagel, Düren, Germany) using nitrogen as the carrier gas (flow rate 0.7 $mL{\cdot}min^{-1}$). For the determination of methanol, the oven temperature gradient profile was 45 °C (4 min) to 180 °C (5 min) at 20 °C/min and a final step of 250 °C (4 min); for acetic acid, it was 45 °C (4 min) to 180 °C (3 min) at 20 °C/min and to 250 °C (6 min) at 30 °C/min. For the release of bound methanol, 1.3 mL of sample (0.75% pectin solution), 100 µL of *n*-propanol (0.1%), and 600 µL of 2 M NaOH were filled into a 10 mL GC vial, sealed, and kept for 1 h at 40 °C. External calibration was done using methanol in a range of 0.01–0.04% (w/v) with *n*-propanol as the internal standard (0.01% w/v). For the release of bound acetic acid, 0.75% pectin solution (2.7 mL) was combined with 900 µL 2 M NaOH in a 5 mL volumetric flask for 1 h at room temperature. Sixty µL propionic acid (1%) and 1110 µL sulphuric acid (1 M) were added to ensure a pH < 2.0. The final reaction mixture (2.5 mL) was used for analysis. External calibration was performed with acetic acid in a range of 0.04−0.12% (w/v) using propionic acid as an internal standard (0.1% w/v). Unsaponified control samples were analyzed with an equal amount of water replacing the NaOH solution. Values of DM and DA were calculated as mole methyl/acetyl groups per 100 mol of GalAc as described earlier.[25]

Molecular size distribution was determined by high-performance size exclusion chromatography (HPSEC) with refractive index (RI) detection.[26] Sample preparation included dialysis against demineralized water (MWCO 3.5 kDa) and filtration through 0.2 µm Chromafil RC-20/15 MS filters (Macherey-Nagel, Düren, Germany). Samples were adjusted to 50 mM sodium nitrate concentration. Eight pullulan standards ranging from 6.1 to 708 kDa (ReadyCal-Kit Pullulan, PSS-Polymer Standards, Mainz, Germany) were used to calculate MWs. The analysis was conducted on a Smartline HPLC system with a RI detector 2300 (Knauer, Berlin, Germany) equipped with two different, connected SEC-Diol columns (300 and 120 Å, 3 μm; YMC, Kyoto, Japan). Samples (20 μL) were injected and eluted with 50 mM sodium nitrate and 0.0025% sodium azide (pH 6.8) for 30 min at a flow rate of 0.3 $mL \cdot min^{-1}$.

The viscosity (mPa·s) of samples was determined using a rotational viscometer (V-Pad, Fungilab, New York City, NY) equipped with an LCP spindle. The analysis was run for 1 min and 200 rpm at 22 °C.

The monomer composition of soluble oligosaccharides was analyzed after acid hydrolysis according to the respective enzyme kit from Megazyme (Wicklow, Ireland). According to the manufacturers' instructions, samples were hydrolyzed with sulfuric acid (2 M) at 100 °C (6 h) for the determination of GalAc and hydrochloric acid (1.3 M) at 100 °C (1 h) for contents of rhamnose and fucose, respectively. The assays are linear over the range of 0.5−100, 5−100, and 5−150 μg for fucose, rhamnose, and GalAc, respectively. The smallest differentiating absorbance corresponds to 0.34, 0.6, and 8.7 $mg \cdot L^{-1}$ for fucose, rhamnose, and GalAC, respectively. The principle of all assays is based on the enzymatic oxidation of the specific sugar by forming nicotinamide-adenine dinucleotide (NADH), which increases stoichiometrically with the amount of specific sugar and was measured photometrically at 340 nm.

Specific monosaccharides were analyzed in the supernatant after centrifugation with a Heraeus Megafuge 40R centrifuge (Thermo Fisher Scientific, Braunschweig, Germany) at 10,947 *g*, 20 min and 10 °C.

Anthocyanin Extraction. Anthocyanins for mixture A (cyanidin glycosides) were extracted from chokeberry juice kindly provided by Haus Rabenhorst O. Lauffs GmbH & Co. KG (Unkel, Germany). Anthocyanins for mixture B (anthocyanidin glucosides) were extracted from purple corn (Nurtisslim d.o.o., Vrhnika, Slovenia) and commercial red wine (Cabernet Sauvignon). A polyphenol-enriched extract was obtained by using column chromatography on Amberlite XAD7 HP (Sigma-Aldrich, Munich, Germany).[27,28]

Further purification of mixture A was performed by membrane chromatography with the membrane adsorber Sartobind S IEX 150 mL (Sartorius Stedim Biotech, Göttingen, Germany).[29] After washing with water to remove sugars and polar compounds, polyphenols including anthocyanins were eluted with ethanol/acetic acid (19:1 v/v) and lyophilized. Further purification of mixture A was performed by membrane chromatography with the membrane adsorber Sartobind S IEX 150 mL (Sartorius Stedim Biotech, Göttingen, Germany).[29] Mixture B was subjected to high-performance countercurrent chromatography (HPCCC) separation [30,31] using a model DE Spectrum centrifuge (Dynamic Extractions, Tredegar, UK). The preparative coil consists of polytetrafluoroethylene (PTFE) tubes (tubing diameter 1.6 mm, a total volume of 136 mL). The HPCCC system was equipped with a Blue Shadow 40P solvent-delivery pump (Knauer, Berlin, Germany), a Blue Shadow D50 UV/vis detector (Knauer, Berlin, Germany), a Foxy R1 fraction collector (Teledyne ISCO, Lincoln, NE), a Degasys DG-1210 degasser (Uniflows, Tokyo, Japan), and a recirculating chiller F-108 (Büchi, Essen, Germany). The separation was run at a revolution speed of 1600 rpm and a flow rate of 6 $mL \cdot min^{-1}$. The solvent system consisted of n-butanol/MTBE/acetonitrile/water (2:2:1:5, v/v/v/v), acidified with 0.1% trifluoroacetic acid (TFA). The organic phase was used as the stationary phase in head-to-tail elution mode. Elution was monitored at 520 nm. Fractions were collected in 30 s intervals.

Red wine and purple corn anthocyanins were combined at a ratio of 125 $mg \cdot L^{-1}$/175 $mg \cdot L^{-1}$ of total anthocyanin content (TAC) in order to obtain a balanced anthocyanin profile only including different anthocyanidin-3-*O*-glucosides.

UHPLC-DAD Quantification of Anthocyanins. Ultrahigh-performance liquid chromatography diode array detector (UHPLC-DAD) analysis was performed on a Prominence UFLC system (Shimadzu, Kyoto, Japan) equipped with two Nexera *X2* LC-30AD high-pressure gradient pumps, a Prominence DGU-20A5R degasser, a Nexera SIL-30AC Prominence autosampler (15 °C, injection volume 5 μL), a CTO-20AC Prominence column oven (40 °C), and a SPD-M20A Prominence diode array detector. Data acquisition and processing were performed using LabSolutions software version 5.85 (Shimadzu, Kyoto, Japan). Anthocyanin separation was carried out on an ACQUITY UPLC HSS T3 column (2.1 μm, 150 × 1.8 μm; Waters, Milford, MA). The following gradient was used at a flow rate of 0.4 $mL\cdot min^{-1}$/(min/% B): 0/4, 2/4, 7/8, 13/10, 20/13, 22/14, 25/19, 26/30, 26.30/100, 28.30/100, 28.80/4, and 32.00/4, where eluent A was water/formic acid (95:5, v/v) and eluent B was acetonitrile/formic acid (95:5, v/v). Anthocyanins were detected at 520 nm and quantified as cyanidin-3-*O*-glucoside (cya-3-glu) equivalents. TAC was calculated as the sum of individual anthocyanins. Analytes were identified by comparing retention times, elution order, and UV/vis spectra with corresponding literature.[32–34]

Incubation of Native Pectin and Pectin Residues with Anthocyanins. Solutions of native SBP, UMP, or EMP were incubated with anthocyanin mixture A or B according to Buchweitz et al. (2012).[35] Prior to incubation, pectin solutions were centrifuged with a Heraeus Megafuge 40R centrifuge (Thermo Fisher Scientific, Braunschweig, Germany) at 10 947 *g*, 20 min, and 10 °C to remove insoluble residues. Concentrations were adjusted to 100 $mg\cdot L^{-1}$ TAC, 0.3% (w/v) pectin and 0.2% (w/v) potassium sorbate for preservation. Anthocyanin concentration was determined after incubation for 2 hours or 2 weeks at 25 °C in an incubator (APT.line KB incubator, Binder GmbH, Tuttlingen, Germany) under gentle shaking. Each experiment was carried out in triplicate. Prior to UHPLC-DAD analysis, samples were centrifuged at 12 500 *g* for 30 min (Heraeus Pico 17, Thermo Fisher Scientific, Braunschweig, Germany) and filtered through 0.2 μm Chromafil RC-20/15 MS filters (Macherey-Nagel, Düren, Germany).

Statistical Analysis. To determine significant differences, an ANOVA with Bonferroni posthoc test was performed using software XLSTAT version 2014.4.06 (Addinsoft, Paris, France). The level of significance was defined as $p \leq 0.05$. Statistical analysis and principal component analysis (PCA) were conducted also using XLSTAT.

3 Results and Discussion

Characterization of Modified Pectin. The MW distribution was determined by HPSEC to obtain information on the extent of degradation of EMP and UMP. Both treatments decreased the average MW compared to native SBP with a molecular peak (Mp) of ~300 kDa (**Figure 3-1**). Distribution patterns of EMP are comparable to Muñoz-Almagro et al.,[36] who degraded apple and citrus pectin by enzyme treatment.

Data for DM and DA is listed in **Table 3-1**. The results of SBP (DM 55–57%, DA 27–31%) are in the range previously reported[35] and were in general concordance with the manufacturer's declaration (66% DM, 32% DA). The amount of dissociated GalAc is supposed to play a key role in the interaction with anthocyanins.[16,21] According to DM, ~45–43% of GalAc groups are free of methylation. The number of dissociated GalAc's (**Table 3-1**) was calculated according to Plaschina et al.,[37] who pointed out that the degree of dissociated GalAc depends on the pK_a value and DM. GalAc, rhamnose, and fucose were determined to examine the monomer composition of SBP, EMP, and UMP (**Table 3-1**). The amount of GalAc can be correlated to the relative amount of HG, while rhamnose and fucose, due to their distinct occurrence in pectin substructures, can be used as markers for RG I and RG II, respectively.[19]

Ultrasound-Induced Modification. Ultrasound treatment decreases MW with increasing treatment time (**Figure 3-1, A**). Ultrasonication provokes two effects that lower MW. Shear forces, generated by the strong motion of solvent and polymers, and cavitation phenomena break chemical bonds within the polysaccharide chains.[9] Additionally, radicals induced by ultrasound react with the polymers and lead to chemical degradation. The narrow MW distribution indicates a small polydispersity index and homogeneity of MW.[9] The results of monomer analysis show the sugar composition of UMP (**Table 3-1**).

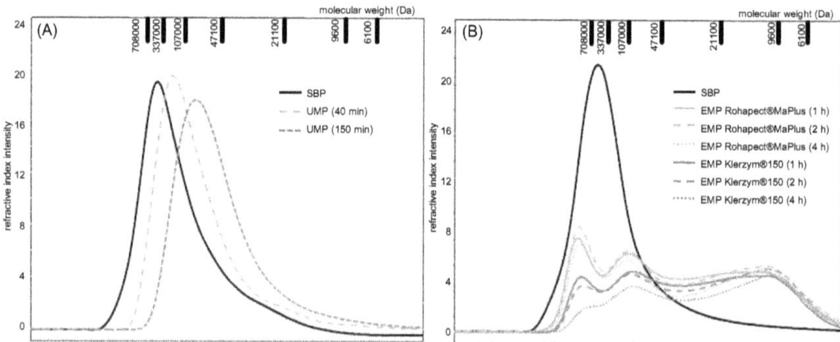

Figure 3-1. HPSEC elution pattern of (A) native SBP and UMP (60% amplitude; 40 and 60 min) and (B) native SBP and EMP (100 ppm: Rohapect®MaPlus 1, 2 and 4 h; Klerzym®150 1, 2 and 4 h). Column was calibrated with standards illustrating the peak MW distribution.

Table 3-1: Characterization of Soluble Pectin after Ultrasound and Enzyme Treatment Compared To Corresponding Untreated Control of Sugar Beet Pectin (SBP). Different Letters Indicate Significant Differences Between Treatment Times

	treatment time	DM (%)	DA (%)	GalAc (*µmol*)	not-esterified GalAc[a] (*µmol*)	dissociated COO^{-a} (*µmol*)	rhamnose (*µmol*)	fucose (*µmol*)
Ultrasonic-modified pectin	SBP 1	57.27 $_a$	31.22 $_a$	8.17 $_a$	3.34	1.54	0.11 $_a$	0.01 $_a$
	40 min	53.07 $_b$	29.60 $_a$	8.25 $_a$	3.73	1.84	0.14 $_b$	0.02 $_a$
	150 min	52.68 $_b$	25.88 $_b$	8.20 $_a$	3.74	1.63	0.17 $_b$	0.04 $_b$
enzyme (Rohapect MaPlus)- modified	SBP 2	54.60 $_a$	27.26 $_a$	8.97 $_a$	3.91	1.77	0.13 $_a$	0.01 $_a$
	1 h	32.35 $_b$	11.20 $_b$	9.07 $_b$	6.04	2.02	0.15 $_{a,b}$	0.01 $_a$
	2 h	31.34 $_c$	6.89 $_c$	9.07 $_b$	6.13	1.99	0.16 $_{a,b}$	0.01 $_a$
	4 h	25.50 $_d$	5.63 $_d$	9.61 $_b$	7.08	2.14	0.18 $_b$	0.01 $_b$
enzyme (Klerzym 150)- modified pectin	SBP 3	55.27 $_a$	28.79 $_a$	9.04 $_a$	3.83	1.71	0.08 $_a$	0.01 $_a$
	1 h	35.70 $_b$	12.37 $_b$	9.67 $_b$	6.10	2.14	0.10 $_{a,b}$	0.02 $_{a,b}$
	2 h	29.68 $_b$	6.98 $_c$	9.67 $_b$	6.70	2.15	0.12 $_{a,b}$	0.02 $_{a,b}$
	4 h	22.79 $_c$	5. 97 $_c$	9.81 $_b$	7.68	2.16	0.16 $_b$	0.02 $_b$

[a]Calculated values depending on pH, DM, pK_a, and degree of dissociation according to Plaschina et al.[37]

The ultrasound treatment did not generate any free monomers (data not shown), which is in agreement with Muñoz-Almagro et al.[36] UMP mainly consisted of GalAc predominantly representing HG polysaccharide residues. Ultrasound did not affect the amount of GalAc, which remained constant compared to SBP. Contents of rhamnose and fucose were low but increased after ultrasound treatment, indicating the release of RG I and RG II residues. In SBP, over 90% of GalAc is part of the HG chain and, accordingly, only a small part of RG I and RG II exists.[38] The results suggest that ultrasound treatment causes a breakdown of the chain and branches of RG-like polymers concurrently with an unaffected HG backbone.[9,36]

The DM and DA of UMP were slightly decreased compared to SBP (**Table 3-1**). Ultrasound reduces DM by either mechanical forces or due to the formation of free radicals.[19] While the decrease in DM caused by ultrasound has often been described, a decrease in DA has not yet been reported.[9,19] Ma et al. noticed no significant change of DA after ultrasonication of pectin. This might be explained by the very low DA of citrus pectin and the relatively mild ultrasound conditions that were applied.[19] The mechanism of the decrease in DA is probably the same as that described for DM.

The viscosity of the treated sample significantly decreased from 5.31 ± 0.03 mPa·s for SBP to 2.81 ± 0.01 mPa·s and 2.54 ± 0.01 mPa·s for 40 and 150 min treatments, respectively. The reduction in viscosity is related to a decrease in MW and DM.[9] The results imply that the voluminous native structure of pectin was modified to smaller, unfolded, and linear polymers, forming a less viscous solution.[19] Ultrasound generates a pool of polysaccharides that consist of shorter HG chains and less branched RG I polymers compared to SBP. The backbone of UMP remains unaltered.[36] Additionally, UMP contained RG I residues, proven by an increased rhamnose content (**Table 3-1**). The increased concentration of fucose in UMP suggests that ultrasound treatment releases the highest amount of RG II-like polymers compared to other samples, which can be seen from the PCA biplot (**Figure 3-2**).

Enzyme-Induced Modification. Enzyme treatments resulted in a broad MW distribution presenting a great polydispersity of generated oligosaccharides in EMP (**Figure 3-1, B**). The two different enzyme preparations resulted in a similar degradation patterns of pectin. Both enzymes substantially decreased the MW with ongoing treatment time. The MW patterns are in agreement with Kressmann, who degraded black currant pectin with enzyme preparations for 1–2 h.[39] Kleryzm150 appeared to be more efficient than RohapectMaPlus in degrading pectin to low MW oligosaccharides at comparable dosage and incubation time. EMP contained two different high-MW fractions (around 400 kDa and 50–90 kDa, respectively) together with a broad range of low-MW oligomers displayed by SEC analysis (**Figure 3-1, B**). The first fraction presumably consists of nondegraded native pectin, and the second may comprise released, highly branched RG-like polymers.[39] Oligomers with MW < 10 kDa, included in the third fraction in SEC distribution pattern, may include RG II dimers or may be counted to the so-called "modified hairy

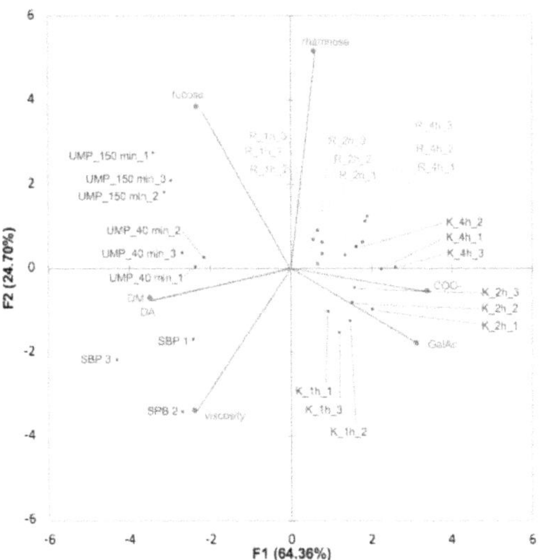

Figure 3-2: PCA biplot of characteristics of SPB, EMP and UMP (n = 3). EMP is categorized according to the enzyme applied: K for Klerzym150 and R for RohapectMaPlus.

regions" (MHRs) described by Hilz et al.[5] These authors point out that most enzyme preparations applied in fruit juice processing are not able to entirely degrade pectin but rather leave soluble oligosaccharides and polysaccharides. They contain RG I attached to arabinan and arabinogalactan side chains of various lengths with HG stubs depending on the side activities of the enzyme preparation.[5] Oligomers with lower MW (<1 kDa) are described as poly-GalAc residues from HG or arabinans and alternating rhamnose-GalAc residues from RG I.[18] Whereas RohapectMaPlus includes only polygalacturonase (PG) and pectin methylesterase (PME), Klerzym150 contains multiple activities including a pectin lyase (PL) and several side activities. PL can break methylated HG chains, while PG needs the preliminary activity of PME. Considering the monomer composition, EMP of Klerzym150 showed a higher amount of GalAc compared to EMP of RohapectMAPlus, and both have significantly higher levels than SBP **(Table 3-1)**. There was no detectable free GalAc after treatments, indicating the absence of *exo*-activities.[39] EMP mainly consists of GalAc and minor quantities of rhamnose and fucose. Klerzym150 released a higher amount of rhamnose (referred to SBP 3) representing RG-degrading activities like rhamnogalacturonase (**Table 3-1** and **Figure 3-2**).[18] RohapectMaPlus revealed a comparable low release of RG-like oligosaccharides (referred to SBP 2) due to the lack of enzyme side activities necessary to degrade hairy regions. The increase in fucose content in oligosaccharides is not significant, but the trend suggests that EMP contains RG II fragments caused by RG II releasing *endo*-polygalacturonases. Results indicate that the monomer composition of both EMPs slightly differs, caused by the variation of enzyme activities of the preparations applied. This effect is displayed by the PCA biplot, where EMP samples of corresponding preparation can be distinguished (**Figure 3-2**). The enzyme composition of Klerzym150 brings along a higher potential to break down smooth and hairy region than RohapectMaPlus, which contains only PG and PME. Additionally, the PCA biplot shows that the two EMPs get more similar within characteristics when treatment time increases (**Figure 3-2**).

EMP showed a decrease in DM and DA up to 53% and 79% for RohapectMaPlus and up to 59% and 80% for Klerzym150, respectively **(Table 3-1)**. Although the acetyl esterase side activity is not declared, such activities are common for fruit-processing enzymes. There are acetyl esterases specific for HG or RG regions, and the latter is essential for RG degradation. After deacetylation of GalAc residues in RG, the branched regions can be degraded further by RG-specific hydrolases and lyases.[40]

The viscosity of EMP strongly decreased by > 50% (5.11 ± 0.11 mPa·s for SBP, 2.18 ± 0.05 mPa·s 4 h for Klerzym150, and 2.28 ± 0.04 mPa·s 4 h for RohapectMaPlus), which proved the breakdown of the pectin to soluble polysaccharides and oligosaccharides. These are characterized by short HG oligosaccharide and polysaccharide residues, less branched RG-like polymers with HG stubs, and a strongly reduced DM and DA compared to SBP and UMP (**Figure 3-2**). Lower DM leads to a high amount of dissociated GalAc residues (COO^-).

PCA analysis was performed to reveal any possible correlations between the different treatments. The biplot (**Figure 3-2**) summarizes the discussed results. According to the PCA biplot, UMP, EMP, and SBP are well-distinguished, enabling a qualitative differentiation of native pectin compared to modified polysaccharides and oligosaccharides and also within modification techniques. The PCA biplot reveals that the main difference between the two enzyme preparations applied is the ability of Klerzym150 to break down hairy regions to RG-like oligosaccharides. In contrast to that, UMP is characterized by higher MW, DM, and DA and a higher amount of RG I compared to EMP.

Pectin Anthocyanin Interactions. SBP, UMP, or EMP were incubated with two different anthocyanin mixtures A or B containing either cyanidin-3-*O*-glycosides (mixture A: arabinoside (ara), galactoside (gal), glucoside (glu), xyloside (xyl)) or anthocyanindin-3-*O*-glucosides (mixture B: delphinidin (del), cyanidin (cya), petunidin (pet), peonidin (peo), malvidin (mal)). Anthocyanin concentration was determined by UHPLC-DAD after 2 h and 2 weeks to differentiate between immediate and long-term effects. The changes in the anthocyanin profile of each sample are displayed in **Figure 3-S1** in the **Supporting Information**. All samples revealed significant alterations in the concentration of each anthocyanin compared to a pectin-free control, confirming an interaction between every studied anthocyanin and all applied pectin types (modified or native); previous reports have only described effects of native pectin changing the concentration of cya, pet and perlagonidin (pel) molecules.[17,20,41] A PCA analysis of SBP and UMP or EMP was performed to verify any correlations between pectin polymers and the type of anthocyanin (**Figure 3-3** and **3-4**).

SBP Anthocyanin Interactions. Native pectin has often been reported to interact with polyphenols, especially anthocyanins. The driving forces of the formation of these complexes are van der Waals interactions, hydrogen bonding, and hydrophobic interactions.[17,21,41] The interaction is strongly influenced by the ultrastructure and structural flexibility of pectin and polyphenols. Native pectin consists of a voluminous

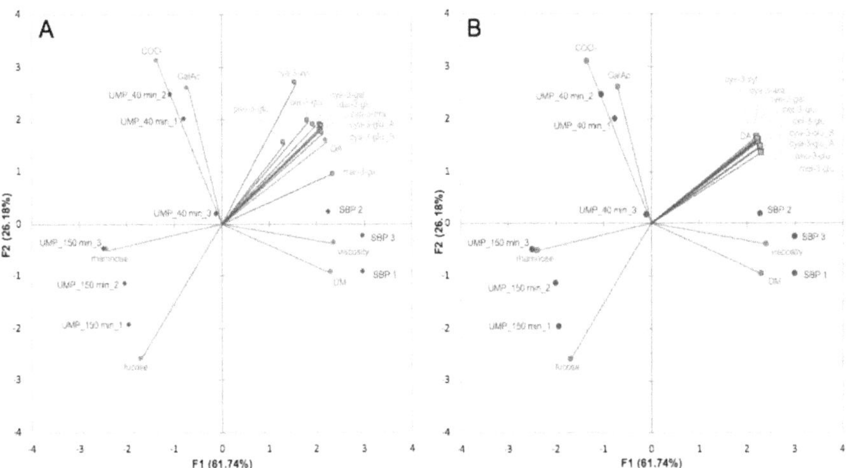

Figure 3-3: PCA biplot of SPB and UMP characteristics with supplementary data of the change of anthocyanins after incubation of (A) 2 h and (B) 2 weeks.

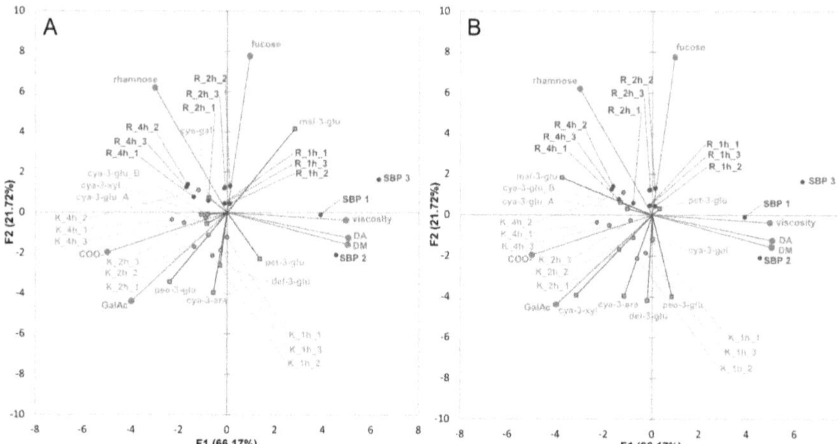

Figure 3-4: PCA biplot of SPB and EMP characteristics with supplementary data of the change of anthocyanins after incubation of (A) 2 h and (B) 2 weeks. EMP is categorized according to the enzyme applied: K for Klerzym150 and R for RohapectMaPlus

porous complex providing hydrophilic domains and hydrophobic pockets where polyphenols can be retained.[42] The interactions toward anthocyanins are mainly governed by hydrogen bonds formed by hydroxy groups of the anthocyanin B-ring and dissociated GalAc in the HG chain.[16,17,20] The slightly acidic pH enhances the interactions by facilitating the formation of ionic interactions between negatively charged GalAc residues and positively charged flavylium cations, which are in equilibrium with their hemiketal form at pH 3.5.[17,43] Only small effects of immediate interactions were observed, shown by a decrease in the TAC of 2% and 6% for mixtures A and B, respectively, although the amount of dissociated GalAc (**Table 3-1)** of SBP (1.54–1.77 μmol·mL^{-1}) provided enough potential binding sites for all anthocyanins (0.56 μmol·mL^{-1}). This decrease was caused by the formation of insoluble anthocyanin–pectin complexes. The entangled structure of native SBP might hide the amount of dissociated GalAc and, thus, sterically hinder the ionic interaction with anthocyanins. In contrast, SBP addition led to a higher TAC up to 7% and 5% for mixture A and B, respectively, after 2 weeks of incubation. The formed complexes can protect anthocyanins against oxidation and other degradation pathways.[21] This effect apparently outweighs the effects described for the immediate interactions, which lead to insoluble complexes. Results indicated that protective soluble complexes had more influence on anthocyanin content than insoluble complexes, because the TAC was higher in SBP solution compared to pectin-free control after long-term storage.

Concentrations of individual anthocyanins were differently affected by SBP addition (**Figure 3-S1**). The concentration of cya-3-xyl and cya-3-glu displayed the greatest alterations by immediate and long-term incubation (–9.3%, +6.9% and –6.0%, +4.6%, respectively) in mixture A compared to a pectin-free control. Several anthocyanins from mixture B were strongly affected by both incubation periods: del-3-glu (–16.9%, +17.1%), pet-3-glu (–15.9%, +3.0%), peo-3-glu (–11.6%, +7.6%), mal-3-glu (–7.1%, +3.5%). Results indicate that the anthocyanidin residue is a more determining factor for complex formation than the sugar residue. The three hydroxy groups on the B-ring explain the reinforced pectin affinity of del, which forms more hydrogen bonds than other anthocyanidins.[16,17] Anthocyanins including methoxy groups on the B-ring probably interact via hydrophobic interactions like van der Waals forces. Both interaction effects outweigh the decrease in TAC caused by immediate interaction to a positive change of TAC after long-term incubation compared to the pectin-free control, demonstrating protective properties.

UMP Anthocyanin Interactions. UMP revealed a significantly stronger ability to reduce anthocyanin content compared to SBP and EMP. The decrease in TAC was 17% and 15% for mixture A and B, respectively, for UMP (150 min) after short-term incubation. UMP readily interacts with anthocyanins to form insoluble complexes, leading to a lower anthocyanin concentration in solution. The decrease can be attributed to the comparable high MW with a homogeneous distribution of UMP polymers (**Figure 3-1A**). UMP decreases TAC up to 70% after long-term incubation. The effect can be explained by an increasing number of pectin–anthocyanin complexes. Additional intermolecular accumulation of anthocyanins, which is called self-association or self-copigmentation, might occur as a side effect to eventually decrease TAC in solution. This mechanism was suggested to be slower and might therefore only play a role during storage.[41,44] It can be assumed that both mechanisms occur, although pectin provides enough binding sites for all anthocyanins. UMP might facilitate anthocyanin stacking more than SBP and EMP because UMP polymers exhibit a linear structure without any hindering side chains or entanglements. Because of these complexations, hydration of anthocyanins might be prevented, which can be considered as a protective mechanism that needs further proof.

UMP polymers consisted mainly of HG chains with dissociated GalAc residues (COO^-) and hydroxy groups accessible on the surface. This is enhanced by reduced DM and DA in UMP compared to SBP. Reduction of DM and DA was more pronounced at longer treatment times (**Table 3-1**). Consequently, UMP contained several exposed groups as potential binding sites for anthocyanins. PCA analysis revealed that DA was positively correlated with the concentration of several anthocyanins (**Figure 3-3**). A reduced DA led to a lower anthocyanin concentration, which might be explained by an increased number of binding sites for complex formation. A low DA reduces steric hindrances by acetyl groups in adjacent positions to potential binding sides.[45] Changes of cya-3-glu and del-3-glu concentrations showed the highest linear correlation ($R^2 < 0.9$) by immediate binding, whereas correlations approximated for all anthocyanins after long-term incubation. No correlation was revealed for the DM, which can be explained by the lower steric hindrance caused by methylation.

UMP contained significantly higher amounts of rhamnose and fucose (**Table 3-1**), indicating soluble RG I and RG II residues. Although the amounts of RG I and RG II residues were comparatively low in the experiments (**Table 3-1**), the results indicate that these residues of the hairy region participate in the interaction with hydrophobic anthocyanins like peo, pet and mal. The contents of these anthocyanins were most affected by the addition of UMP (150 min, **Figure 3-S1**). While the complexation of the flavanone hesperidin that bears comparable substituents on the B-ring has been reported to interact with neutral sugars of the hairy region,[46] the hydrophobic interaction to anthocyanins has not been studied so far.

Individual anthocyanins showed different binding affinities toward UMP polymers by immediate and long-term interactions (**Figure 3-S1**). Concentrations of cya-3-glu and cya-3-xyl from mixture A and del-3-glu and peo-3-glu from mixture B are most affected by immediate interactions with UMP, which is similar to the effects of SBP interactions. The concentrations of cya-3-gal, cya-3-ara, and del-3-glu were strongly reduced by complex formation after long-term incubation. It can be assumed that the sugar moieties induce steric hindrances of the interactions and therefore influencing the complexation of individual anthocyanins.

EMP Anthocyanin Interactions. EMP caused a similar decrease in TAC compared to SBP by immediate interactions. The effect did not significantly change the anthocyanin profile compared to SBP **(Figure 3-S1)**. Additionally, the PCA biplot revealed no correlation between anthocyanin type and pectin characteristics (**Figure 3-4A**). SEC analysis showed that EMP still contained nondegraded SBP residues **(Figure 3-1B)** that might affect anthocyanin content similar to the effects described for SBP.

In contrast, anthocyanin content was differently affected by EMP compared to SBP after long-term incubation (**Figure 3-S1**). Both EMP and SBP interactions with anthocyanins revealed a protective effect. Whereas the anthocyanin concentration of mixture A did not significantly change compared to SBP, the concentrations of anthocyanin glucosides from mixture B were positively affected and resulted in higher TAC than SBP (5.4% and 6.0–10.1% SBP and EMP, respectively). The results indicate that both the type of enzyme and treatment time that is correlated with the extent of degradation of pectin affected pectin characteristics and consequently the anthocyanin profile. EMP generated by Klerzym150 (2 and 4 h) and RohapectMA Plus EMP (4 h) preserved the highest contents of del-3-glu (**Figure 3-S1**). Prolonged enzyme treatment led to decreasing DM and increasing amount of dissociated GalAc, generating more potential binding sites. Holzwarth et al. showed that pectin modified by only PME or PG improved anthocyanins stability of pel and cya molecules after 4 weeks of storage compared to untreated pectin. Pectin treated by PME led to much higher retention of anthocyanins, due to liberated GalAc binding sites.[47]

EMP also preserved a higher content of the anthocyanin-*O*-glucosides of pet, peo, and mal compared to SBP. These anthocyanins include methoxy groups on the B-ring that interact via hydrophobic interactions with pectin rather than via hydrogen bonds. Pet has two hydroxy and one methoxy group that can interact in both ways. Native pectin includes hydrophobic domains in the ultrastructure,[48] which might be preserved after enzyme degradation in RG residues and HG residues with methylation and acetylation. The latter was considerably decreased by PME activities and acetyl esterase activities (**Table 3-1)**. Consequently, soluble residues of RG complex anthocyanins containing methoxy groups on their B-ring, as mentioned earlier, lead to a protective effect. Mixture B contains 36% anthocyanins with methoxy groups (pet, peo and mal). Consequently, an amount of 0.20 $\mu mol \cdot mL^{-1}$ of these anthocyanins can interact via hydrophobic interactions to an equivalent amount of RG-like polysaccharides (**Table 3-1**).

The findings suggest that retention of a broad anthocyanin profile like in bilberries including both hydrophilic and hydrophobic anthocyanins[29] requires various pectin polymers with hydrophilic and hydrophobic domains. Previous studies are mainly focused on hydrophilic anthocyanins, e.g., from strawberries,[16,21] and their interactions. Because reports on the hydrophobic interactions are scarce, their impact might generally be underestimated.

All results demonstrate that ultrasound and enzyme treatment of SBP resulted in a pool of polysaccharides and oligosaccharides with lower MW comprising different MW distribution patterns and a decreased DA and DM. Results indicate that treatments degrade smooth and hairy region of pectin differently. Consequently, SBP, UMP, and EMP affected anthocyanin concentration and profile individually by creating insoluble or soluble complexes. The findings suggest that the binding mechanism depends on hydrophilic and hydrophobic interactions because all studied anthocyanins were affected by pectin. SBP contained a sufficiently high number of dissociated GalAc binding sites to interact with anthocyanins via hydrogen bonds. However, the results demonstrated that this effect is not the sole determining factor for the interaction as frequently been reported[16,17]. UMP included linear HG polymers, less branched RG residues, and accordingly the released RG branches. DM and DA were only slightly reduced by ultrasonication. The polymers high in MW formed insoluble anthocyanin–pectin complexes, intensely lowering the TAC, whereby individual anthocyanins showed different interactions at immediate and long-term incubation with the formed polysaccharides. Sugar residue and aglycone moiety of anthocyanins both influenced complexation. EMP contained polysaccharides and oligosaccharides of a broad range of MWs, considerably reduced in DM and DA. The latter led to 20% more dissociated GalAc residues. These oligosaccharides include HG and RG residues low in MW that can form soluble complexes with anthocyanins, which might result in stabilizing effects.[5,47] Results indicated that both HG-like and RG-like oligosaccharides play a role in complex formation with anthocyanins, which has not been studied so far. The former contains many hydroxy groups for hydrogen bonds to polar anthocyanins with more hydroxy groups. Less polar anthocyanins bearing methoxy groups on their B-ring might interact with hydrophobic domains preserved in RG residues.

The complexation effects between modified pectin and anthocyanins examined in this study might play a crucial role during the maceration step of juice production. The results of this study indicate that the interaction of these compounds might depend on the solubility of complexes and, therefore, influence the concentration of anthocyanins in juice. Although these results are based on model solutions containing SBP and anthocyanin extracts, it can be assumed that fragments of berry pectin, which are formed during juice production, interact analogously with the berry anthocyanins. Because these effects govern the quality of color and the potential health-promoting properties of juices, representing two criteria with growing consumer interest, juice producers need a deeper understanding of these interactions. Because different enzyme preparations comprise numerous pectinase activities, yielding distinct pectin residues, the choice of the enzyme during fruit juice processing is essential for the conservation of the native anthocyanin profile. More knowledge of the impact of enzyme treatment is needed to optimize juice yield along with a high content of anthocyanins, resulting in a high-quality product. A deeper understanding of the mechanism of anthocyanin-pectin interactions can lead to enhanced processing methods that increase the content and stability of anthocyanins.

Funding sources

The project is supported by funds of the Federal Ministry of Food and Agriculture (BMEL) based on a decision of the Parliament of the Federal Republic of Germany via the Federal Office for Agriculture and Food (BLE) under the innovation support program (281A102716).

Supporting information

Figure 3-S1. Change of anthocyanin's profile after (1) 2 h and (2) 14 days of incubation with native, enzyme modified pectin (EMP) and ultrasonic modified pectin (UMP) considering a pectin-free control to subtract all non-pectin relating effects. A: extract A including cyanidins-*O*-glycoside; B: extract B including anthocyanin-*O*-glucosides.

EMP 1A

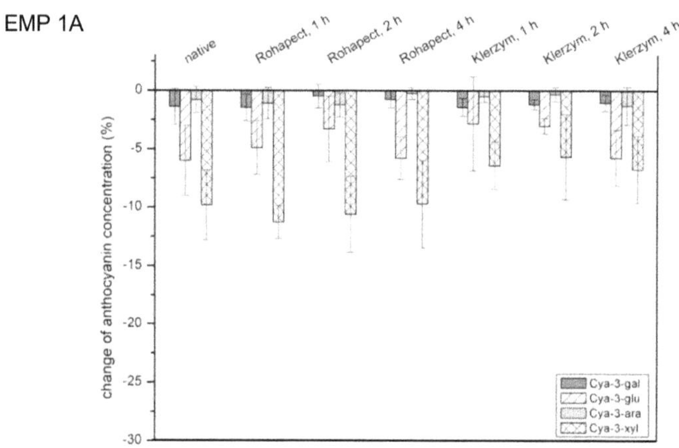

EMP 1B

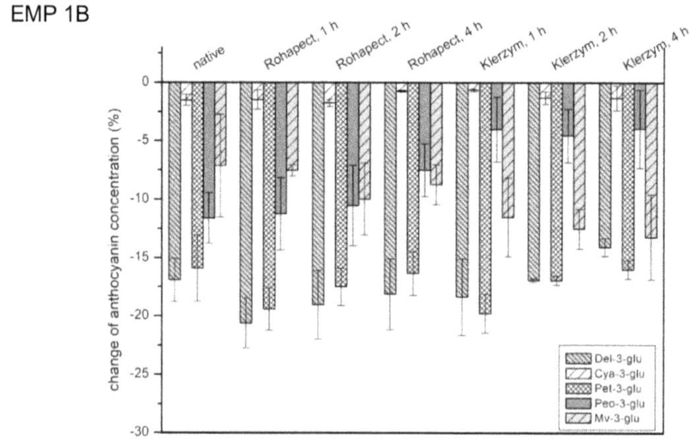

EMP 2A

Cya-3-gal
Cya-3-glu
Cya-3-ara
Cya-3-xyl
change of anthocyanin concentration (%)
native
Rohapect, 1 h
Rohapect, 2 h
Rohapect, 4 h
Klerzym, 1 h
Klerzym, 2 h
Klerzym, 4 h

EMP 2B

Del-3-glu
Cya-3-glu
Pet-3-glu
Peo-3-glu
Mv-3-glu
change of anthocyanin concentration (%)
native
Rohapect, 1 h
Rohapect, 2 h
Rohapect, 4 h
Klerzym, 1 h
Klerzym, 2 h
Klerzym, 4 h

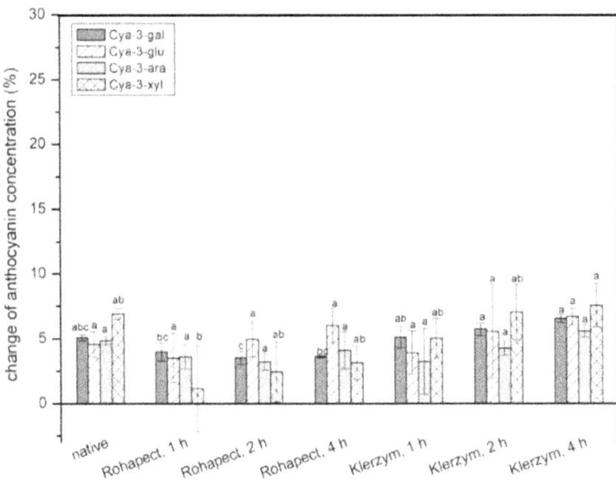

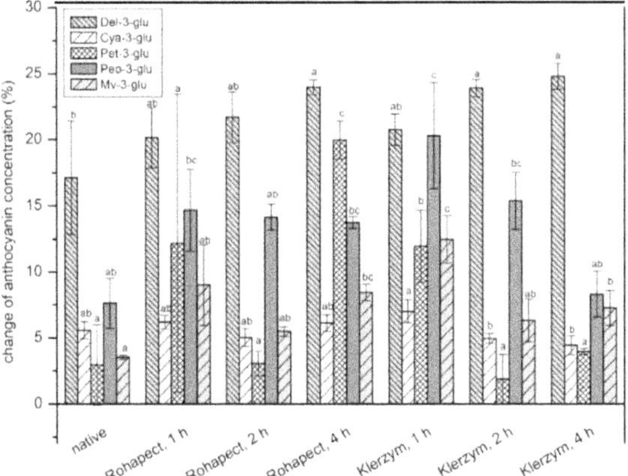

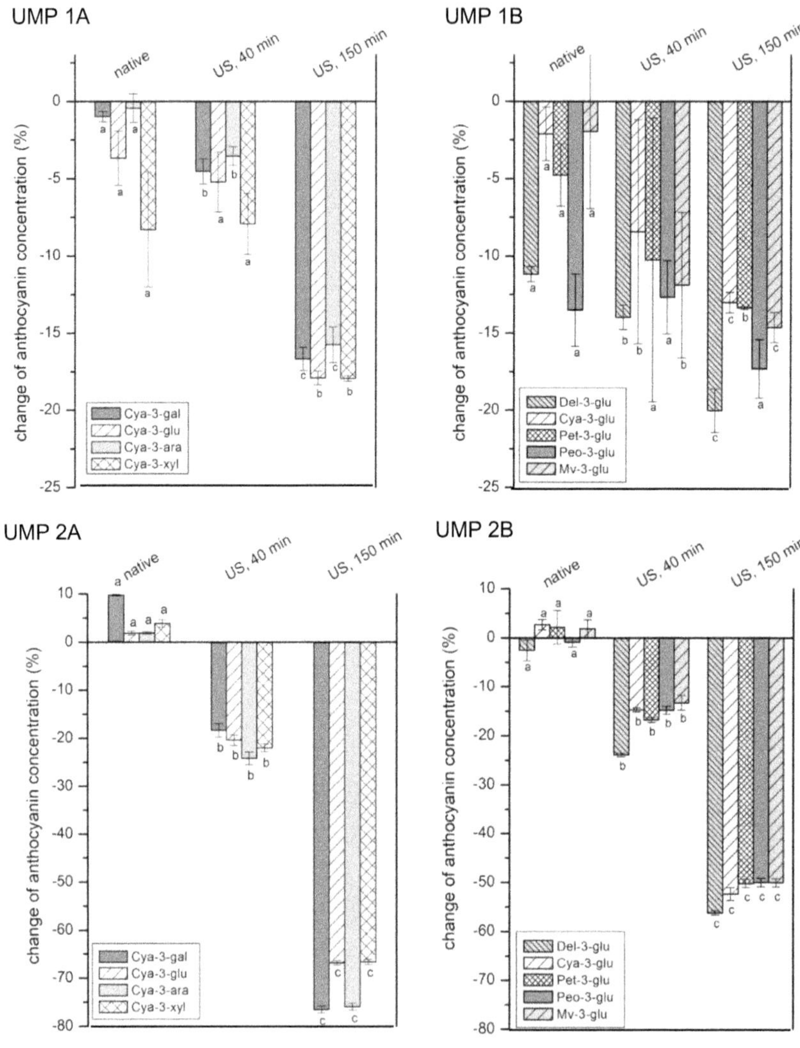
UMP 1A
native
US, 40 min
US, 150 min
change of anthocyanin concentration (%)
Cya-3-gal
Cya-3-glu
Cya-3-ara
Cya-3-xyl
UMP 1B
Del-3-glu
Cya-3-glu
Pet-3-glu
Peo-3-glu
Mv-3-glu
UMP 2A
UMP 2B

References

(1) Mazza, G.; Kay, C. D. *Bioactivity, absorption, and metabolism of anthocyanins*; Blackwell Publishing: Oxford, UK, 2008; Vol. 1.

(2) Seeram, N. P. Recent trends and advances in berry health benefits research. *J. Agric. Food Chem.* **2010**, *58* (7), 3869–3870.

(3) Weber, F.; Larsen, L. R. Influence of fruit juice processing on anthocyanin stability. *Food Res. Int.* **2017**, *100*, 354–365.

(4) White, B. L.; Howard, L. R.; Prior, R. L. Impact of different stages of juice processing on the anthocyanin, flavonol, and procyanidin contents of cranberries. *J. Agric. Food Chem.* **2011**, *59* (9), 4692–4698.

(5) Hilz, H.; Bakx, E. J.; Schols, H. A.; Voragen, A. Cell wall polysaccharides in black currants and bilberries - characterisation in berries, juice, and press cake. *Carbohydr. Polym.* **2005**, *59* (4), 477–488.

(6) Hartmann, A.; Patz, C. D.; Andlauer, W.; Dietrich, H.; Ludwig, M. Influence of processing on quality parameters of strawberries. *J. Agric. Food Chem.* **2008**, *56* (20), 9484–9489.

(7) Landbo, A. K.; Meyer, A. S. Effects of different enzymatic maceration treatments on enhancement of anthocyanins and other phenolics in black currant juice. *Innov. Food Sci. Emerg. Technol.* **2004**, *5* (4), 503–513.

(8) Oszmiański, J.; WojdyŁo, A.; Kolniak, J. Effect of enzymatic mash treatment and storage on phenolic composition, antioxidant activity, and turbidity of cloudy apple juice. *J. Agric. Food Chem.* **2009**, *57* (15), 7078–7085.

(9) Zhang, L.; Ye, X.; Ding, T.; Sun, X.; Xu, Y.; Liu, D. Ultrasound effects on the degradation kinetics, structure and rheological properties of apple pectin. *Ultrason. Sonochem.* **2013**, *20* (1), 222–231.

(10) Patras, A.; Brunton, N. P.; O'Donnell, C.; Tiwari, B. K. Effect of thermal processing on anthocyanin stability in foods; mechanisms and kinetics of degradation. *Trends Food Sci. Technol.* **2010**, *21* (1), 3–11.

(11) Voragen, A. G. J.; Coenen, G. J.; Verhoef, R. P.; Schols, H. A. Pectin, a versatile polysaccharide present in plant cell walls. *Struct. Chem.* **2009**, *20* (2), 263–275.

(12) Ridley, B. L.; Neill, M. A. O.; Mohnen, D. Pectins: Structure, biosynthesis, and oligogalacturonide-related signaling. *Phytochemistry* **2001**, *57* (6), 929–967.

(13) Øbro, J.; Harholt, J.; Scheller, H. V.; Orfila, C. Rhamnogalacturonan I in *Solanum tuberosum* tubers contains complex arabinogalactan structures. *Phytochemistry.* **2004**, *65* (10), 1429–1438.

(14) Hilz, H.; Williams, P.; Doco, T.; Schols, H. A.; Voragen, A. The pectic polysaccharide rhamnogalacturonan II is present as a dimer in pectic populations of bilberries and black currants in muro and in juice. *Carbohydr. Polym.* **2006**, *65* (4), 521–528.

(15) Pinelo, M.; Arnous, A.; Meyer, A. S. Upgrading of grape skins: Significance of plant cell-wall structural components and extraction techniques for phenol release. *Trends Food Sci. Technol.* **2006**, *17* (11), 579–590.

(16) Buchweitz, M.; Speth, M.; Kammerer, D. R.; Carle, R. Stabilisation of strawberry (*Fragaria x ananassa* Duch.) anthocyanins by different pectins. *Food Chem.* **2013**, *141* (3), 2998–3006.

(17) Fernandes, A.; Bras, N. F.; Mateus, N.; de Freitas, V. Understanding the molecular mechanism of anthocyanin binding to pectin. *Langmuir* **2014**, *30*, 8516–8527.

(18) Ducasse, M. A.; Williams, P.; Canal-Llauberes, R. M.; Mazerolles, G.; Cheynier, V.; Doco, T. Effect of macerating enzymes on the oligosaccharide profiles of merlot red wines. *J. Agric. Food Chem.* **2011**, *59* (12), 6558–6567.

(19) Ma, X.; Wang, W.; Wang, D.; Ding, T.; Ye, X.; Liu, D. Degradation kinetics and structural characteristics of pectin under simultaneous sonochemical-enzymatic functions. *Carbohydr. Polym.* **2016**, *154*, 176–185.

(20) Buchweitz, M.; Speth, M.; Kammerer, D. R.; Carle, R. Impact of pectin type on the storage stability of black currant (*Ribes nigrum* L.) anthocyanins in pectic model solutions. *Food Chem.* **2013**, *139* (1–4), 1168–1178.

(21) Holzwarth, M.; Korhummel, S.; Siekmann, T.; Carle, R.; Kammerer, D. R. Influence of different pectins, process and storage conditions on anthocyanin and colour retention in strawberry jams and spreads. *LWT - Food Sci. Technol.* **2013**, *52* (2), 131–138.

(22) Blumenkrantz, N.; Asboe-Hansen, G. New method for quantitativ determination of uronic acids. *Anal. Biochem.* **1973**, *54*, 484–489.

(23) Nunes, C.; Rocha, S. M.; Saraiva, J.; Coimbra, M. A. Simple and solvent-free methodology for simultaneous quantification of methanol and acetic acid content of plant polysaccharides based on headspace solid phase microextraction-gas chromatography. *Carbohydr. Polym.* **2006**, *64*, 306–311.

(24) Savary, B. J.; Nuñez, A. Gas chromatography – mass spectrometry method for determining the methanol and acetic acid contents of pectin using headspace solid-phase microextraction and stable isotope dilution. *J. Chromatogr. A* **2003**, *1017* (1017), 151–159.

(25) Levigne, S.; Thomas, M.; Ralet, M.; Quemener, B.; Thibault, J. Determination of the degrees of methylation and acetylation of pectins using a C18 column and internal standards. *Food Hydrocoll.* **2002**, *16*, 547–550.

(26) Houben, K.; Jolie, R. P.; Fraeye, I.; Van Loey, A. M.; Hendrickx, M. E. Comparative study of the cell wall composition of broccoli, carrot, and tomato: Structural characterization of the extractable pectins and hemicelluloses. *Carbohydr. Res.* **2011**, *346* (9), 1105–1111.

(27) Weber, F.; Greve, K.; Durner, D.; Fischer, U.; Winterhalter, P. Sensory and chemical characterization of phenolic polymers from red wine obtained by gel permeation chromatography. *Am. J. Enol. Vitic.* **2013**, *64* (1), 15–25.

(28) Weber, F.; Boch, K.; Schieber, A. Influence of copigmentation on the stability of spray dried anthocyanins from blackberry. *LWT - Food Sci. Technol.* **2017**, *75*, 72–77.

(29) Juadjur, A.; Winterhalter, P. Development of a novel adsorptive membrane chromatographic method for the fractionation of polyphenols from bilberry. *J. Agric. Food Chem.* **2012**, *60*, 2427–2433.

(30) Degenhardt, A.; Knapp, H.; Winterhalter, P. Separation and purification of anthocyanins by high-speed countercurrent chromatography and screening for antioxidant activity. *J. Agric. Food Chem.* **2000**, *48* (2), 338–343.

(31) Rentzsch, M.; Schwarz, M.; Winterhalter, P. Pyranoanthocyanins - an overview on structures, occurrence, and pathways of formation. *Trends Food Sci. Technol.* **2007**, *18* (10), 526–534.

(32) Garcia-Herrera, P.; Pérez-Rodríguez, M.-L.; Aguilera-Delgado, T.; Labari-Reyes, M.-J.; Olmedilla-Alonso, B.; Camara, M.; de Pascual-Teresa, S. Anthocyanin profile of red fruits and black carrot juices, purees and concentrates by HPLC-DAD-ESI/MS-QTOF. *Int. J. Food Sci. Technol.* **2016**, *51* (10), 2290–2300.

(33) Vergara, C.; Mardones, C.; Hermosín-gutiérrez, I.; Baer, D. Von. Comparison of high-performance liquid chromatography separation of red wine anthocyanins on a mixed-mode ion-exchange reversed-phase and on a reversed-phase column. *J. Chromatogr. A* **2010**, *1217* (36), 5710–5717.

(34) Wu, X.; Gu, L.; Prior, R. L.; McKay, S. Characterization of anthocyanins and proanthocyanidins in some cultivars of *Ribes*, *Aronia*, and *Sambucus* and their antioxidant capacity. *J. Agric. Food Chem.* **2004**, *52* (26), 7846–7856.

(35) Buchweitz, M.; Nagel, A.; Carle, R.; Kammerer, D. R. Characterisation of sugar beet pectin fractions providing enhanced stability of anthocyanin-based natural blue food colourants. *Food Chem.* **2012**, *132* (4), 1971–1979.

(36) Muñoz-Almagro, N.; Montilla, A.; Moreno, F. J.; Villamiel, M. Modification of citrus and apple pectin by power ultrasound: Effects of acid and enzymatic treatment. *Ultrason. Sonochem.* **2017**, *38*, 807–819.

(37) Plaschina, I. G.; Braudo, E. E.; Tolstoguzov, V. B. Circular-dichroism studies of pectin solutions. *Carbohydr. Res.* **1978**, *60*, 1–8.

(38) Ralet, M. C.; Cabrera, J. C.; Bonnin, E.; Quéméner, B.; Hellìn, P.; Thibault, J. F. Mapping sugar beet pectin acetylation pattern. *Phytochemistry* **2005**, *66* (15), 1832–1843.

(39) Kressmann, R. The chemical composition of soluble colloids in the juice of black currants and their effect on processing technology (doctoral dissertation), Justus-Liebig University of Gießen, 2001.

(40) Benen, J. A. E.; von Alebeek, G.-J. W. M.; Voragen, A. G. J.; Visser, J. Pectic esterases. In *Handbook of food enzymology*; Whitaker, J. R., Voragen, A. G. J., Wong, D. W. S., Eds.; Marcel Dekker, In: New York, 2003; pp. 849–856.

(41) Padayachee, A.; Netzel, G.; Netzel, M.; Day, L.; Zabaras, D.; Mikkelsen, D.; Gidley, M. J. Binding of polyphenols to plant cell wall analogues - part 1: Anthocyanins. *Food Chem.* **2012**, *134* (1), 155–161.

(42) Le Bourvellec, C.; Guyot, S.; Renard, C. M. G. C. Non-covalent interaction between procyanidins and apple cell wall material: Part I. Effect of some environmental parameters. *Biochim. Biophys. Acta - Gen. Subj.* **2004**, *1672* (3), 192–202.

(43) Castaneda-Ovando, A.; Pacheco-Hernandez, M. de L.; Paez-Hernandez, M. E.; Rodriguez, J. A.; Galan-Vidal, C. A. Chemical studies of anthocyanins: A review. *Food Chem.* **2009**, *113* (4), 859–871.

(44) Renard, C. M. G. C.; Baron, A.; Guyot, S.; Drilleau, J. F. Interactions between apple cell walls and native apple polyphenols: Quantification and some consequences. *Int. J. Biol. Macromol.* **2001**, *29* (2), 115–125.

(45) Ralet, M. C.; Crépeau, M. J.; Buchholt, H. C.; Thibault, J. F. Polyelectrolyte behaviour and calcium binding properties of sugar beet pectins differing in their degrees of methylation and acetylation. *Biochem. Eng. J.* **2003**, *16* (2), 191–201.

(46) Ben-Shalom, N.; Pinto, R. Natural Colloidal Particles: The mechanism of the specific interaction between hesperidin and pectin. *Carbohydr. Polym.* **1999**, *38* (2), 179–182.

(47) Holzwarth, M.; Korhummel, S.; Carle, R.; Kammerer, D. R. Impact of enzymatic mash maceration and storage on anthocyanin and color retention of pasteurized strawberry purées. *Eur. Food Res. Technol.* **2012**, *234* (2), 207–222.

(48) Le Bourvellec, C.; Bouchet, B.; Renard, C. Non-covalent interaction between procyanidins and apple cell wall material. Part III. Study on model polysaccharides. *Biochim. Biophys. Acta* **2005**, *1725*, 10–18.

Chapter 4

Concluding remarks

During red juice production, the degree of cell wall rupture and the succeeding rate of polysaccharide degradation tremendously determine juice yield, the extraction of plant compounds, filtration issues, haze formation, and shelf life; thereby defining the overall juice quality. Today, tailormade enzyme preparations are used to fulfill the efficient cell wall breakdown during the mashing process. Mainly pectinolytic enzymes cleave cell wall pectic polysaccharides and attached molecules, resulting in considerably reduced viscosities. Although the addition of enzymes brought a tremendous advantage for the juice industry, high amounts of unchanged cell wall material rich in secondary plant compounds remain in the pomace. Additionally, enzymatic maceration lasts for approximately 1-2 hours and requires the mash to be at the temperature optima of enzymes at approximately 50 °C. Both these factors, together, lead to the degradation of oxidatively and thermally sensitive compounds like anthocyanins. To face these drawbacks, the implementation of ultrasound technology into the enzymatic maceration of red berry mash is discussed and investigated in this thesis. For this purpose, a combined application, referred to as ultrasound-assisted enzymatic maceration (UAEM), was conducted on a pectin model solution under reduced temperature (30 °C) and compared to the benchmark enzymatic degradation at standard maceration temperature of 50 °C (**Chapter 2**). The quality of pectin degradation was characterized considering the structural properties and compositions of soluble pectic polysaccharides essential for protective anthocyanin complexations.

High yields of juice revealed by improved cell wall degradation are accompanied by increased concentrations of polysaccharide and oligosaccharide residues. Besides the organoleptic properties of haze and cloudiness, taste and mouth feeling, or the nutritional value of increasing fiber content, these matrix compounds form complexes with polyphenols. In particular, complexation effects toward anthocyanins are of great interest in red juices since they contribute to the key quality of red juices, which is the appealing color and nutritional value. Therefore, current scientific knowledge of complexation mechanisms is discussed and examined in this thesis. In **Chapter 3,** complexation effects between pectin fragments and anthocyanins that can equally arise during red juice processing were examined. For this purpose, pectin was modified by enzymatic and ultrasound treatments and incubated with a wide variety of anthocyanins naturally present in berries. The corresponding results contribute to the profound understanding of the complexation mechanisms providing a basis for the usage of these complexes with beneficial bio- and techno-functional properties.

1 Ultrasound-assisted enzymatic maceration

Although promising evidence like enhanced polysaccharide extraction and degradation within shorter treatment time, accelerated enzyme activities, and high anthocyanin stabilities have been published within different objectives for ultrasound applications, research on ultrasound-assisted optimized red juice production, in particular, is in its early infancy. So far, juice-related research has mainly focused on juice yield, anthocyanin concentrations, antioxidant activity, or color changes, revealing improved results in juice quality compared to conventional maceration treatments. In fact, red berry mashes contain several other matrix compounds, in particular, diverse polysaccharides from complex cell wall structures, numerous polyphenols, and few proteins that need to be considered in juice production. Regarding distinct process-induced effects, basic research focusing on individual compounds is essential to understand the underlying mechanisms and reactions. Therefore, the present thesis focuses on the effects of UAEM on pectin degradation, one of the essential requirements during maceration and juice production. For this purpose, a pectin model solution was treated with a commercial enzyme preparation for red juices assisted by different ultrasound treatments, varying in intensities and duration (**Chapter 2**). Exclusive and combined treatments were conducted to distinguish between specific and synergistic effects. The pectin model solution (pH 3.5) was chosen to avoid interfering effects of other berry mash matrix compounds, which would hamper the investigations.

Comparing the present study results of the exclusive ultrasound and enzymatic degradation, the resulting pool of pectic polymers significantly differed in their overall structural properties, encompassing molecular weight, composition, and resulting structure. In detail, ultrasound application reduced the molecular weight of pectic polysaccharides and formed polymers uniform in their molecular weight distribution. Due to the comparably low intensities applied (<50 W/ml), the average molecular weight was reduced by one-third but was still above 100 kDa. In contrast, enzymatic cleavage revealed a diverse pool of oligosaccharides and polysaccharides with wide molecular weight distribution from 5 to 700 kDa.

While smaller polysaccharides are required for juice production due to processing issues (filtration and pressing) and assumed to increase juice quality (as discussed later), exclusive ultrasound treatments alone do not meet the requirements of the desired profile of degraded and released polysaccharides and oligosaccharides. Higher ultrasound intensities are needed to achieve a more pronounced molecular weight reduction. Indeed, increased intensities concomitantly cause conceivable detriments regarding the red juice quality, like uncontrolled temperature, the enlarged formation of reactive radicals, and the corrosion of the ultrasonic probe made of titanium alloy, all of these causing undesired side effects. Correspondingly, the application of enzymes for an appropriate pectin degradation remains necessary for fruit juice production. In fact, the implementation of lower ultrasound intensities enhances and accelerates enzymatic maceration by synergistic effects in a gentle way that finally provides the novel note to improve juice quality.

As the present study investigates, the combined application of UAEM is able to improve pectin degradation into smaller soluble polysaccharides and oligosaccharides that are considered to improve juice quality in several aspects. First, they enrich juices in fiber content accompanied by various biological activities and metabolic functions that were even shown to be enhanced by ultrasound modification. Recently, studies reported *in vitro* and *in vivo* antioxidant, antitumor, anticoagulant, anti-inflammatory, and prebiotic activities of various ultrasound-treated polysaccharides. Smaller polysaccharides are assumed to be more easily transported into cells and metabolized. Also, bacteria, e.g., *Lactobacilli* and *Bifidobacteria* from the gut, easily utilized polysaccharides with reduced viscosities that were produced after ultrasound treatments (Cui and Zhu 2021). Second, distinct polysaccharides bind polyphenols like anthocyanins forming soluble complexes with stabilizing effects, which contributes to nutritional value and color quality of juices, as elaborated in **Section 2**. Additionally, soluble polymers with higher surface activity and smaller molecular weight enhance interfacial capacity and, thus, tend to improve emulsion stability, contributing to haze stability in juices (Bai et al. 2017; Cui and Zhu 2021).

Detailed degradation mechanisms of pectin by UAEM are part of recent investigations and in the case of pectin examined (**Chapter 2**) and discussed in this dissertation. Ultrasound forces have been demonstrated to beneficially affect three targets participating in the enzymatic reaction: substrate, enzymes, and the reaction system (Wang et al. 2018). The elevation of static pressure, hydrodynamic stress, torque, and cavitation-induced microstreaming produces shear forces that modify polysaccharides and enzymes and improve mass transfer, decreasing diffusion barriers that accelerate enzyme-substrate interactions. Correspondingly, activation energies (E_a) were remarkably reduced as determined by the linear fit of the Arrhenius plot during ultrasound treatments. Also, Michaelis-Menten kinetics were used to analyze a decrease in energy barriers necessary for reaction by the maximum rate V_{max} under saturated substrate concentration and Michaelis constant K_m indicating the enhanced affinity (Nadar and Rathod 2017).

1.1 Mechanisms and opportunities of the UAEM-induced synergistic effects

Depending on the applied intensity and target structure, ultrasound causes different chemical changes which include depolymerization and the cleavage of cross linkages such as weak hydrogen bridges, ionic interactions, and van der Waals bonds. These alterations influence structure dependent reactions and the corresponding reaction kinetics.

In the case of pectic polysaccharides, the present study shows that the applied ultrasound intensities reduced their molecular weight by the irreversible chain breakage of the backbone and side chains. Thus, residues from the main pectic subunits of HG, RG I, and RG II, were released as soluble residues. Depending on the polysaccharide structure, it is assumed that linear polymers present in the smooth region of HG are cleaved in the middle, while branched polymers from the subunits of the hairy regions, including RG I and RG II are mainly randomly cut on the side chains (Cui and Zhu 2021). In summary, ultrasound treatments modified pectin to simpler unbranched polymers appearing with uniform molecular weights. The highly branched pectic polysaccharides of the native molecule are barely accessible and thus resistant to common enzymatic degradation. In contrast, ultrasound-modified polymers are more easily accessible for enzymes. The resulting increased surface areas without branching barriers facilitate enzymatic binding and cleavage, thus contributing to the synergistic effect.

Like the polysaccharide structure, enzyme conformations are also modified by ultrasound forces. The low-intensities applied produce stable cavitation bubbles rather than the extremely imploding ones. Stable cavitation induces oscillatory forces that change the conformation of enzymes gently without denaturation. The alterations of enzymes in their secondary and tertiary structure and active sites by ultrasound treatments have been investigated by spectroscopic techniques such as circular dichroism (CD) spectroscopy, intrinsic fluorescence spectroscopy, and Fourier-transform infrared spectroscopy (FTIR). Ultrasound-induced effects have been described to unfold protein molecules by slight conformational changes that lead to an increase in the surface with enhanced exposure of active sites (Wang et al. 2018). These alterations, which may finally improve substrate affinity and facilitate the reaction, probably account for the most significant synergistic effect. In general, the conformation changes strongly influence enzyme catalytic activity, stability, and selectivity.

However, the effects of lower intensity ultrasound do not necessarily induce enhanced enzyme activities, as demonstrated in the present study. Pectinolytic enzyme activities contained in the applied preparation were distinctly affected during ultrasound treatments in both ways, enhanced or decelerated. This fact can be explained by their individual appearances formed by different conformation types (such as α-/β-helices, α-/β-sheets, and turns) and intramolecular bindings (e.g., disulfide bonds). These have been shown to be diversely affected by ultrasound treatments resulting in different structural alterations. Enzymes tend to be inactivated with increasing intensities because harsher ultrasound conditions lead to the destruction of their conformation accompanied by a loss of their structural-depended active domains. Next to their activity defined by the respective product concentration, ultrasound treatment may also prolong their reaction ability over time. It can be assumed that there is only a thin line of the appropriate ultrasound energy input to support each specific activity of the enzyme.

Today, only a few enzymes have been analyzed regarding their structural and activity alterations under ultrasound treatments (Nadar and Rathod 2017; Wang et al. 2018). Many studies focused on inactivation purposes and did not consider lower ultrasound intensities to improve activity rates (O'Donnell et al. 2010). Briefly, published results are quite inconsistent and difficult to compare since experimental conditions vary significantly. Yet, considering ultrasound-induced effects on enzyme specific conformation and the corresponding influences that it has on enzyme activity, it may be a beneficial technique to control enzyme activities during juice maceration in both ways, for enhancement or denaturation purposes. Sufficiently long activities by enhanced enzyme stability provide a beneficial and sustainable approach for an efficient cell wall degradation during juice maceration, allowing lower dosages with reducing expenses. Precise denaturation allows also a fast pasteurization alternative avoiding long degradative effects of heat on other vulnerable and valuable compounds. In particular, it would be interesting to conduct further investigations on appropriate ultrasound power inputs, controlled by frequencies, intensities, and duration to reveal either favorable conformation changes in proteins aiming at highly active and stable enzymatic activities or destructive changes achieving enzyme inactivation when desired.

In the present study, activity rates and product concentrations of the three main pectinolytic enzymes have been considered under ultrasonication, which are polygalacturonase (PG), pectin lyase (PL), and pectin methylesterase (PME). In fact, pectinolytic enzyme preparations for juice maceration contain additional side activities necessary to break down the complex pectic structure of the cell wall network. While the knowledge of individual ultrasound effects on each enzyme is essential to understand ultrasound-induced effects on degradation mechanisms, cell wall degradation should also be considered with all acting enzymes. Like PME demethylates galacturonic acid residues and, thus, provides the substrates for PG, other synergistic relations between enzyme activities can be expected.

Next to the understanding of ultrasound-controlled enzyme activities, the controlled production of the resulting polymers regarding quantity and quality should be considered since these act as bio- and techno-functional compounds in the juices. The present work demonstrated that the applied novel technique of UAEM significantly alters the profile of the arising polymers compared to the exclusive enzymatic treatment. Thus, further investigations need to focus on the characterization of these polymers regarding solubility, molecular weight, monosaccharide composition, and resulting structure since these determine their bio- and techno-functional properties.

It needs to be considered that the present work only reveals pectinolytic degradation in a model solution. The cell wall matrix appears in a highly dense complex which is specific for each fruit and berry type as described above. Thus, further investigations need to focus on UAEM degradation effects on other cell wall compounds like cellulose, hemicellulose, and finally, the total cell wall matrix. It would be interesting to conduct further research considering, e.g., short high-intensity ultrasound pretreatments to lose the dense network of cell wall structures before adding enzymes at lower intensity UAEM treatments, thereby accelerating enzyme efficiency.

Besides the process effects on cell wall polysaccharides, other matrix compounds like proteins or polyphenols need to be considered to expand the knowledge of the overall process-induced effects during juice production. Here, not only direct process-induced modifications or degradation need to be studied, but also subsequent reactions of modified molecules are important of consideration, as the essential complexation of modified polysaccharides and anthocyanins.

Figure 4-1 illustrates the different ways of pectic degradation based on the mechanistic investigations of the present thesis into polysaccharides complexed with anthocyanins during production of red juices.

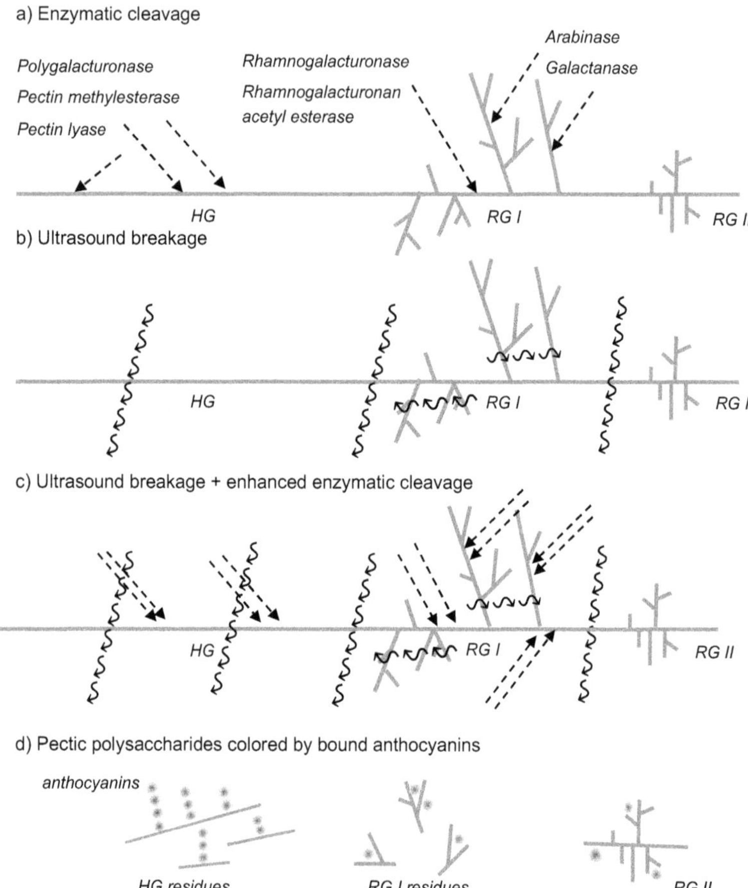

Figure 4-1: Schematic representation of pectin degradation investigated in the present dissertation by a) the cleavage by different enzymatic activities present in enzyme preparations for fruit juice maceration, b) ultrasound treatment, c) ultrasound-assisted enzymatically enhanced or decelerated activities. d) Resulting polymers of different pectic subunits after ultrasound-assisted degradation interact with anthocyanins (represented by small globes) to colored complexes. Linear HG residues provide stacking of anthocyanins by co-pigmentation effect. Dashed and waved arrows indicate the specific point of cleavage or breakage. Homogalacturonan (HG), rhamnogalacturonan (RG).

1.2 Temperature and industrial up-scaling issues

In the present study, the UAEM results demonstrated a comparable or even enhanced process output in pectin degradation under reduced temperature conditions (30 °C) compared to the standard juice maceration temperature of approximately 50 °C, considering enzyme optima. A temperature reduction is beneficial for heat-sensitive compounds during the maceration process and bears energy-saving potential. The reduction has also been demonstrated to be more efficient for the ultrasound-induced synergistic effects on enzyme activities and extraction issues. Previous studies are in agreement with unpublished data of pretrials, noting that most ultrasound-induced effects occur at temperatures significantly below the enzyme temperature optima. In detail, pretriails (data not shown) did not illustrate significant differences when enzyme (PG and PL) reactions were carried out close to their temperature optima (40-55 °C) with and without ultrasound treatments. Plausible reasons are not clear yet. It is assumed that under reduced temperatures local cavitation-induced hotspots raise temperature locally by reaching enzyme optima shortly, while the medium temperature remains constantly low. At higher medium temperatures, these local hotspots might lead to heat-induced denaturation rather than beneficial effects. Since tailormade enzymes are quite efficient for their purposes, reactions under optimum temperature occur very fast, and ultrasound support probably becomes negligible or acts adversely. This does not imply any redundancy of ultrasound addition since the combined application degrades polysaccharides efficiently, resulting in a significantly changed polymer profile. Additionally, ultrasound addition allows a temperature reduction during processing, which is beneficial for all heat-sensitive compounds, thus improving juice quality.

Indeed, all enzyme activities should be screened for their performance under ultrasound treatments in further studies. As shown in the present study, the enzyme pectin acetyl esterase did not reach a comparable activity at reduced temperatures compared to the benchmark results at its optimum temperature (50 °C), which was determined by the lower reduction of the degree of acetylation. Once again, each enzyme reacts individually under ultrasound treatments enormously depending on both reaction temperature and energy inputs.

Also, ultrasound-assisted extraction of plant material has been shown to achieve higher yields by low-temperature treatments. At higher temperatures, air inclusion forming bubbles accelerates, and corresponding collapses produce less intense shock waves due to a cushioning effect, both caused by lower liquid viscosities and surface tensions. In contrast, at lower temperatures, smaller bubbles are formed and collapse with relatively high intensities shock waves, causing cell wall rupture and mass transfer (Cui and Zhu 2021). Thus, ultrasound effects are greater at lower temperatures. Subsequently, ultrasound power reduction should be considered to avoid harsh cavitational conditions regarding sensitive compounds like enzymes and polyphenols. Additionally, the reduction provides energy-saving potential for both parameters, temperature and ultrasound.

In conclusion, temperature plays a crucial role in UAEM and ultrasound tremendously increases temperature. Correspondingly sample temperature control is important during the whole process and requires appropriate cooling systems. In the present study, a continuous system was used that maintained constant temperature over the entire treatment. The ultrasonic probe was fixed in a flow cell with counterflow cooling. Additionally, the sample was tempered in a water bath (**Figure 2-1)**. In contrast, several studies treated samples batch-wise in flasks by only ice bath cooling to keep temperatures low, but with questionable outcome. Instead of sample temperature, mostly the temperature of the cooling media was reported. Thus, high-temperature effects cannot be excluded in these studies.

Besides adequate temperature management, the use of a continuous system including an implemented ultrasound probe into a flow cell provides several other advantages. Power inputs can be controlled more easily because of the short contact times of the samples with the ultrasound probe. Systems are scalable to industrial purposes, where continuous systems are strongly preferred for production lines. Additional ultrasound probes can be integrated or easily exchanged along a production line to treat large amounts of mash efficiently.

A drawback of the ultrasound technique worth mentioning is the vulnerable ultrasound probe material made of titanium alloy, which is likely to corrode during high-intensity energy transmission, especially in acidic environments like berry mashes. The erosion may reduce the efficiency of ultrasound energy transmission and probe shelf life. The possible migration of metal traces into the juices implies food safety and sensory concerns. Therefore, approaches of a non-contact process have already been considered to overcome this drawback. Here, liquid intermedia transmit ultrasound waves into the food product that is separated from the sonoreactor (Feng et al. 2011; Freitas et al. 2006).

Although ultrasound systems are already offered for food production, experience and industrial scale research is lacking, hampering companies to invest in the novel technology. Additionally, a lack of proper documentation of parameters used in previous research makes the usage of ultrasound techniques at an industry level complicated (Delgado-Povedano and Luque de Castro 2015). Parameters of ultrasound processing conditions should clearly be described in each study so that comparative studies can be performed to maximize research efficiency and efforts. Overall, ultrasound-induced effects depend on numerous factors such as ultrasound devices and parameters of total energy input controlled by frequency and intensity, processing time, medium-temperature and volume, and finally, the characteristics of the target compound. Published data are already promising to improve juice quality and change production in a sustainable manner. Therefore, it would be of great interest to conduct further research using ultrasound-assisted enzymatic maceration, especially on industry scale.

2 Anthocyanin-pectin complexation in red berry juices: a perspective at the molecular level

The greater the cell wall degradation, the greater is the amount of extracted polyphenols, especially anthocyanins during mashing. Subsequently, polyphenols, and polysaccharides that are mainly pectin-derived, interact with each other throughout production and storage. The consequences are multiple; polyphenol associations with food macromolecules have been shown to significantly impact organoleptic and nutritional qualities. Several effects of polyphenol-polysaccharide complexation have been examined in wide-ranging areas, e.g., tannins in vinification (Riou et al. 2002), proanthocyanidins in apple juices (Le Bourvellec et al. 2004, 2005), anthocyanins in jams (Holzwarth et al. 2013), anthocyanins in coloring food additives (Buchweitz et al. 2012), as well as bioavailability issues *in vitro* and *in vivo* (Mazza 2007). The underlying complexation mechanisms are diverse and not fully understood so far. The present work brought new insights into the interaction mechanisms between anthocyanins and pectin. The corresponding findings are discussed regarding the partly diverse and conflicting results of studies dealing with polyphenol-polysaccharide binding mechanisms. In **Section 3.2**, complexes are considered in terms of fruit juice-related aspects.

2.1 Mechanisms of polyphenol-polysaccharide interactions

In general, interactions between cell wall polysaccharides and polyphenols are described as non-covalent interactions such as van der Waal forces, hydrophobic interactions, and hydrogen bonds low in energy and largely reversible. The type and number of possible bonds are strongly influenced by the characteristics of both polyphenols and polysaccharides. It is still challenging to explicitly rank the various factors that drive the interactions because they might be synergistic or antagonistic in the same system (Liu et al. 2020).

In the specific case of anthocyanin-polysaccharide interactions, three different binding mechanisms, *i.e.,* ionic interactions, hydrogen bonds, and stacking by self-association, are currently discussed in literature, with contradicting conclusions about the predominating effect. The findings of the present study (**Chapter 3**) contribute to a deeper understanding of these interaction mechanisms where different binding types are involved and explain the observed outcomes. In short, structurally different pectic polymers, varying in their molecular weight, monosaccharide composition, and spatial structure, were incubated with anthocyanins varying in their aglycone type and sugar moiety to extend the knowledge of interaction dependencies.

Probably the strongest interaction is provided by ionic forces between the anthocyanin cation and negatively charged dissociated galacturonic acid residues in acidic environments (**Figure 1-3**). Here, an increased number of dissociated carboxylic residues in the pectin backbone seems to facilitate the electrostatic interactions. However, at low pH (<2.0), where anthocyanins occur as flavylium cation, pectin carboxy groups are protonated due to the pK_a of galacturonic acid (3.5). Experiments in the present study were conducted at pH 3.5, considering the average acidity of the berries. At this pH, ionic structures of both counterparts are likely present. Since the dissociation rate of galacturonic acid in pectin depends also on the degree of methylation (Plaschina et al.

1978), approximately 20-25% of the non-esterified galacturonic residues are dissociated, as demonstrated in **Chapter 3.** Additionally, the ionic binding might possibly cause a shift in anthocyanin equilibrium towards the flavylium cation since the charged form is selectively stabilized by its interaction with pectic polysaccharides and thus withdrawn from the equilibrium.

Regarding polysaccharide structures, linear and galacturonic acid-enriched polymers (represented by UMP in **Chapter 3**) revealed the highest anthocyanin binding concentrations. Similarly, Fernandes et al. (2021) examined in their study that distinct pectic polysaccharides extracted from grapes rich in HG and poor in neutral side chains revealed also higher anthocyanin binding capacities compared to pectic grape polysaccharides rich in neutral branched regions. Analogous results have been shown for the binding mechanism of procyanidins, which have a dimeric structure and require even more space compared to the smaller anthocyanins (Watrelot et al. 2014). The presence of neutral side chains probably hinders polyphenol complexation through a more compact structure and steric hindrances. In **Chapter 3**, the ultrasound-modified pectic polymers (UMP) appeared to possess reduced hydrodynamic volumes that facilitated accessibility towards anthocyanins forming insoluble complexes.

Besides the ionic interaction, hydrogen bonds are supposed to play a key role in the complexation. Thus, polygalacturonic acid chains display many hydroxy and carboxy groups providing an appropriate surface for hydrogen bonds towards anthocyanins (**Figure 1-3**). In the anthocyanin structure, the hydroxy groups in the B-ring have been demonstrated to predominantly participate in hydrogen bonds. Results of **Chapter 3** and several studies confirmed these findings for both counterparts (Buchweitz et al. 2013; Fernandes et al. 2015). Accordingly, pectic polysaccharides with higher numbers of free hydroxy or carboxy groups increased their binding affinity towards anthocyanins. Here, a low degree of methylation and acetylation contributes to the interaction by more free galacturonic acid residues. On the other hand, the anthocyanin delphinidin (three hydroxy groups) showed greater binding affinities than cyanidin (two hydroxy groups) due to the higher hydrogen binding potential by the B-ring hydroxy groups. This fact has also been shown in other studies (Buchweitz et al. 2013; Fernandes et al. 2015). However, the influence of the B-ring hydroxy groups might only account for anthocyanins. Other flavonoids have been demonstrated to yield contradicting results considering only the B-ring hydroxy groups. For example, luteolin (two hydroxy groups) show higher adsorption affinities towards oat β-glucan than quercetin (two hydroxy groups) and myricetin (three hydroxy groups) (Wang et al. 2013). Conceivably, hydroxy groups in the A- or C-rings might also play a role in the interaction mechanisms. Additionally, the precise location of hydroxy groups has been shown to affect the adsorption differently (Wang et al. 2013; Zhu et al. 2018).

Although it appears that in the case of anthocyanins, the B-ring is involved most in hydrogen bonds, significant interactions have been shown between the terminal rhamnose moiety of neohesperidin dihydrochalcone and cyclodextrin by intermolecular hydrogen bonds (Caccia et al. 1998; Malpezzi et al. 2004). Different degrees of glycosylation between flavonoids may lead to differences in both solubility and steric hindrance, which then turn into changes in adsorption affinities (Liu et al. 2020). Thus, it would be interesting to study other naturally abundant anthocyanins containing more than just one glycoside residue. Concluding, further investigations are needed to evaluate and rank the significance of hydrogen bonds and, in particular, their locations for the effects on the polyphenol-polysaccharide interactions in general.

In the present study, anthocyanin adsorption increased remarkably over two weeks, especially in the case of linear UMPs (**Chapter 3**). This phenomenon can be explained by the binding mechanism of self-association, where anthocyanins do not bind directly to polysaccharides but associate with their aromatic rings due to π–π stacking. The stacking appearance in the presence of polysaccharides was first published by Padayachee et al. (2012), who distinguished between an initial anthocyanin binding on polymers, followed by slower anthocyanin parallel stacking, eventually accounting for the major amount of bound anthocyanins. The high binding concentrations by linear polymers can be explained by less steric hindrances by branching or entanglements. Likewise, linear polymers of cellulose and xyloglucan showed a higher rate of saturation caused by their conformations favoring stacking, whereas branched polymers revealed only low binding concentrations of anthocyanins (Le Bourvellec et al. 2005). The driving forces of anthocyanin self-association are described by London dispersion forces, as part of van der Waals forces, and hydrophobic interactions. The former contributes dominantly to π-π-interactions between two non-polar aromatic molecules by induced dipoles. Hydrophobic interactions are based on the exclusion of water molecules between two approaching hydrophobic interfaces, resulting in entropy gain (Trouillas et al. 2016).

An increase in hydroxy or methoxy groups in the anthocyanin B-ring has been demonstrated to enhance self-association. Methoxy groups tend to influence the interaction more strongly than the hydroxy groups (Hoshino 1991; Trouillas et al. 2016). The electron-rich substituents extend the π-electron system beneficially for π-π-stacking. Methoxy groups have a less pronounced mesomeric effect (+M) compared to hydroxy groups, resulting in a more uniform electron contribution over the molecule that might favor non-covalent dispersion forces. Additionally, the less polar methoxy groups cause an enhancement in the hydrophobic interactions in comparison to the polar hydroxy groups. Furthermore, substituents are known to influence the torsion of the B-ring and differentially affect the planar alignment favoring the vertical stacking in chiral aggregates (Houghton et al. 2021).

Nevertheless, the parallel stacking of several molecules requires an alternation of electron donor and electron acceptor (Trouillas et al. 2016). While the methoxy group might form an electron acceptor system, the hydroxy groups probably act as the electron donor, respectively, due to the higher mesomeric effect of the hydroxy group (+M). This consideration might explain why all anthocyanins examined in the present study were retained by complexation after long-term incubation. The different number and location of hydroxy and methoxy groups in the molecules form anthocyanins with slightly diverse electrostatic distributions caused by the different mesomeric effects of the substitutions. Thus, stacking might be favored by an alternating arrangement of all these anthocyanins. This consideration should be tested experimentally in the future, where stacking ability is compared between similar and structurally different anthocyanins. Related studies have only recently been published, illustrating the current relevance of the understanding of the anthocyanin binding mechanisms in the presence of pectic polysaccharides (Fernandes et al. 2021; Koh et al. 2020; Liu et al. 2020).

The discussed interaction mechanisms, ionic binding, and stacking, are facilitated by one distinct pH-dependent anthocyanin structure of either the flavylium cation (pH >2) or the quinoidal form (pH 3.5), respectively. The ionic binding requires the oppositely charged molecules of pectin residues and the flavylium cation. Although stacking occurs for both anthocyanin structures favored by their ring-planar alignment, the repulsive forces of the similarly charged flavylium cations hamper the self-association (Fernandes et al. 2015). Thus, stacking probably increases at higher pH, where the equilibrium shifts to the non-polar quinoidal base. Accordingly, Liu et al. (2020) observed that the interaction of cyanidin-3-glucoside and pectin was nearly doubled at pH 3.6 compared to lower values at pH 2.0. These findings indicate that not only is stacking favored at the higher pH, but also a remarkable anthocyanin complexation ability is noticed. Indeed, the primarily bound flavylium cations on the pectin surface might be neutralized and, thus, favor stacking. In the present study, stacking probably accounts for the significantly increased amount in bound anthocyanins after long-term incubation favored at pH 3.5. However, berries may differ in their pH values, which might affect interaction and reduce protective complexation during juice production.

Neutral side chains of branched pectin subunits have been less considered in literature regarding anthocyanin interactions so far. However, they are highly important for hydrophobic interactions toward less polar polyphenols. In the present study, native pectin and especially enzyme-modified pectin (EMP in **Chapter 3**) composed of higher amounts of neutral monosaccharides revealed significant protective effects on anthocyanin concentration after long time incubation as soluble complexes. The results indicate that probably hydrophobic interactions might play a significant role in the binding mechanisms of anthocyanins and polysaccharides with respect to protective properties. Similarly, highly condensed and hydrophobic pectic structures have been observed to bind more anthocyanins through hydrophobic interactions than hydrophilic arabinoxylan or xyloglucan (Padayachee et al. 2012; Phan et al. 2017). Liu et al. (2020) describe two types of hydrophobic interactions between polyphenols and polysaccharides. In the first, polysaccharides encapsulate polyphenols in hydrophobic pockets. In the second, polysaccharides with exposed hydrophobic residues bind to less soluble polyphenols by concomitantly increasing polyphenol solubility (Liu et al. 2020). While the first might account for native pectin, the second type explains the binding of EMP that gained the greatest anthocyanin complexation in the present study (**Chapter 3**). The effects of neutral side chain branching have only been studied for procyanidins, where no differences were observed in the case of a low degree of polymerization (<9), while molecules of higher degrees (>30) impeded interactions (Watrelot et al. 2014). Also, pectin and cyclodextrins have been incubated with different polyphenols, resulting in interactions explained by polysaccharide hydrophobicities, conformational flexibility, and the size of both polysaccharides and polyphenols (Le Bourvellec and Renard 2012). In fact, the results of the present study suggest that a decrease in hydrodynamic volume of the pectic structure, meaning the reduction of molecular weight and side branches, very likely facilitates the binding ability of anthocyanin molecules to the polymer.

As part of another approach, the thermodynamic parameters of the interaction between malvidin-3-glucoside and different pectic polysaccharide fractions extracted from grapes were examined by Fernandes et al. (2021). All interactions revealed negative Gibbs Energies (ΔG), representing spontaneous reactions. Interestingly, the pectic fraction rich in HG domains and with only low amounts in neutral side chains had the highest association constant, inferring the tightest bonds. All interactions revealed negative enthalpy contributions, which indicates that these exothermic interactions are related to the formation of electrostatic interactions and hydrogen bonds. Indeed, entropy contributions increased most, implying that hydrophobic effects might predominantly drive the interactions. The authors explained their findings by favored self-associations that lead to the significant entropy gain by hydrophobic interactions (Fernandes et al. 2021). However, hydrophobic interfaces are also displayed by the side branches of neutral monosaccharides. Accordingly, it was hypothesized in **Chapter 3** that predominantly less polar anthocyanins like peonidin, petunidin, and malvidin, due to their methoxy residues, might interact with neutral residues of the hairy region by hydrophobic interactions. So far, studies on anthocyanin complexation lack plant pectic polysaccharides rich in hairy regions and neutral sugars to explain the findings on the molecular level. Thus, with regard to berry juice production, it would be interesting to examine anthocyanin interactions using pectic polysaccharides of the respective fruits.

Underlined by the summarized findings that are partly conflicting, it is still uncertain whether hydrophilic, hydrophobic, ionic, or dispersion forces predominantly drive the interaction for each counterpart. Probably a combination of all these forces might determine the final effect. It has been clearly demonstrated that the arising forces are highly individual regarding the diverse structural properties of both counterparts, polysaccharides and anthocyanins. Additionally, the surrounding pH has been shown to play a crucial role in binding mechanisms as it determines anthocyanin structure and galacturonic acid dissociation rates. Since all discussed aspects influence the binding affinity by changing the thermodynamic energies, the underlying mechanisms are difficult to discriminate. In fact, the decrease in hydrodynamic volume of the pectic structure very likely facilitates the binding ability towards anthocyanin and can be used for desired complexation questions. During juice production, a controlled pectin modification into smaller and simpler pectic polymers bearing high complexation affinities can protect high amounts of anthocyanins and other polyphenols against degradative reactions, thus increasing juice quality.

2.2 Complexation during red juice production

While in the experimental approaches, the applied polysaccharides are characterized to elucidate the binding mechanisms, during juice production, a large number of undetermined reactive polysaccharides with different structural properties are formed by the already discussed food processing techniques and distinct process conditions like temperature, pressure, and pH. Moreover, each fruit differs in its cell wall composition, affecting extractability, degradation of plant matrix, and finally, the polymer properties (Ortega-Regules et al. 2006). Consequently, with respect to juice production, extracted cell wall polysaccharides vary in their solubility and appear in different molecular weights, spatial structure, sugar compositions, and charge that finally tremendously increase their complexation properties. As examined above, the major cell wall compound pectin is especially susceptible to enzymatic, physical, and chemical modifications during processing, resulting in remarkable depolymerization and demethylation (Liu et al. 2020).

Studies focusing on polyphenol-polysaccharide interactions are scarce, particularly with respect to the processing effects on polysaccharides and their subsequent influence on polyphenol complexation. For this purpose, pectin was modified before complexation experiments with different anthocyanin extracts in the present study by similar maceration conditions (**Chapter 3**). Accordingly, the interaction of modified pectin fragments and individual anthocyanins imitated production-like complexation. Pectin was modified by enzyme and ultrasound treatments, resulting in highly diverse pectin solutions with even greater different complexation potential. The principal component analyses (PCA) performed illustrate distinct correlations of certain polymers characteristics and individual anthocyanins (**Figures 3-3 and 3-4**). Linear polysaccharides formed insoluble complexes with all anthocyanins eluding them from analytical detection. Entangled polysaccharides, especially after enzymatic modification, formed soluble aggregates with reversible bindings revealing protective properties in solution. The results clearly demonstrate that the process-induced effects tremendously modify polysaccharides and increases their interaction potential. The complexes differ in their physicochemical properties, particularly their solubility.

After the conventional enzymatic maceration, colloids remaining in juices have been identified as modified hairy regions (MHR) consisting of strongly branched pectin residues. It needs to be considered that the present work examined reactions with sugar beet pectin, which is structurally different from berry pectin. As explained above, the structural properties of pectin strongly depend on complexation ability. However, regarding the results of the present study and the literature, it can be concluded that all pectic subunits are involved in anthocyanin interactions but pectic polymers reduced in their size and spatial structure might reveal the greatest effects (Fernandes et al. 2020, 2021; Koh et al. 2020; Lin et al. 2016).

According to the results presented in this thesis, the introduced UAEM application provides the potential to produce desired soluble polysaccharides of linear shape and less branches, which might be superior for anthocyanin complexation. This assumption has to be proven by studies applying UAEM to actual red berry maceration. More specifically, it would be interesting to ascertain how to produce soluble polysaccharides by cell wall degradation that protects the majority of the anthocyanins from degradation.

While the present study showed that soluble complexes significantly protected anthocyanins in the supernatant during long-term storage, improving their stability against degradative reactions at 4 °C, other studies further demonstrated complex thermostability properties in ranges of 60-100 °C (Fernandes et al. 2020). Juice production, e.g., during blanching or conventual pasteurization, is accompanied by heat loads destructive for heat-sensitive compounds. Interestingly, different pectic subunits enabled diverse protective properties at different temperatures; in comparison, HG-rich extracts revealed the greatest effects at 60 °C, RG-rich extracts the most at 100 °C (Fernandes et al. 2020). The findings again highlight the different underlying binding mechanisms between different pectic polysaccharides and anthocyanins, bearing different binding strengths. In fact, in several studies, high temperature reported to disrupt hydrogen bonds and reduce polyphenol adsorption toward polysaccharides, such as the interaction between cyanidin-3-glucoside and ferulic acid toward cellulose (Phan et al. 2016), procyanidins with apple cell walls (Le Bourvellec et al. 2005), and tea polyphenols with β-glucans (Wu et al. 2011), respectively. The divergent findings also indicate, that different polysaccharide targets probably possess diverse binding abilities, resulting in different protective potential.

Considering the promising potential of polysaccharides as protective molecules, further studies should focus on appropriate polysaccharide modifications by considering the resulting physicochemical properties, complex stability during processing and storage, organoleptic consequences, and finally, bioaccessibility and bioavailability. An increased depth of knowledge with respect to polysaccharide-polyphenol interactions for food processing is necessary and therefore makes for a very interesting and upcoming area of research. Understanding the driving mechanisms of polysaccharides and polyphenols may allow conclusions of the relation of juice processing and bioavailability that could finally lead to the development of new recommendations to produce healthier and more nutritious juices.

References

Bai, L., Huan, S., Li, Z., & McClements, D. J. (2017). Comparison of emulsifying properties of food-grade polysaccharides in oil-in-water emulsions: Gum arabic, beet pectin, and corn fiber gum. *Food Hydrocolloids, 66*, 144–153.

Buchweitz, M., Nagel, A., Carle, R., & Kammerer, D. R. (2012). Characterisation of sugar beet pectin fractions providing enhanced stability of anthocyanin-based natural blue food colourants. *Food Chemistry*, *132*(4), 1971–1979.

Buchweitz, M., Speth, M., Kammerer, D. R., & Carle, R. (2013). Stabilisation of strawberry (*Fragaria x ananassa* Duch.) anthocyanins by different pectins. *Food Chemistry*, *141*(3), 2998–3006.

Caccia, F., Dispenza, R., Fronza, G., Fuganti, C., Malpezzi, L., & Mele, A. (1998). Structure of neohesperidin dihydrochalcone/β-cyclodextrin inclusion complex: NMR, MS, and X-ray spectroscopic investigation. *Journal of Agricultural and Food Chemistry*, *46*(4), 1500–1505.

Cui, R., & Zhu, F. (2021). Ultrasound modified polysaccharides: A review of structure, physicochemical properties, biological activities and food applications. *Trends in Food Science and Technology*, *107*, 491-508.

Delgado-Povedano, M. M., & Luque de Castro, M. D. (2015). A review on enzyme and ultrasound: A controversial but fruitful relationship. *Analytica Chimica Acta*, *889*, 1–21.

Feng, H., Barbosa-Cánovas, G. V., & Weiss, J. (Eds.). (2011). *Ultrasound Technologies for Food and Bioprocessing*. (Vol 1, p.599). New York, NY: Springer.

Fernandes, A., Brandão, E., Raposo, F., Maricato, É., Oliveira, J., Mateus, N., ... de Freitas, V. (2020). Impact of grape pectic polysaccharides on anthocyanins thermostability. *Carbohydrate Polymers*, *239*, 116240.

Fernandes, A., Brás, N. F., Mateus, N., & de Freitas, V. (2015). Correction to "Understanding the molecular mechanism of anthocyanin binding to pectin." *Langmuir*, *31*(5), 1866–1866.

Fernandes, A., Raposo, F., Evtuguin, D. V., Fonseca, F., Ferreira-da-Silva, F., Mateus, N., ... de Freitas, V. (2021). Grape pectic polysaccharides stabilization of anthocyanins red colour: Mechanistic insights. *Carbohydrate Polymers*, *255*, 117432.

Freitas, S., Hielscher, G., Merkle, H. P., & Gander, B. (2006). Continuous contact- and contamination-free ultrasonic emulsification – A useful tool for pharmaceutical development and production. *Ultrasonics Sonochemistry* *13*(1), 76-85.

Holzwarth, M., Korhummel, S., Siekmann, T., Carle, R., & Kammerer, D. R. (2013). Influence of different pectins, process and storage conditions on anthocyanin and colour retention in strawberry jams and spreads. *LWT - Food Science and Technology*, *52*(2), 131–138.

Hoshino, T. (1991). Self-association of flavylium cations of anthocyanidin 3,5-diglucosides studied by circular dichroism and H-NMR. *Phytochemistry, 31*(2), 647–653.

Houghton, A., Appelhagen, I., & Martin, C. (2021). Natural blues: Structure meets function in anthocyanins. *Plants, 10*(4), 726.

Koh, J., Xu, Z., & Wicker, L. (2020). Blueberry pectin and increased anthocyanins stability under *in vitro* digestion. *Food Chemistry, 302*, 125343.

Le Bourvellec, C., Bouchet, B., & Renard, C. (2005). Non-covalent interaction between procyanidins and apple cell wall material. Part III. Study on model polysaccharides. *Biochimica et Biophysica Acta, 1725(1)*, 10–18.

Le Bourvellec, C., Guyot, S., & Renard, C. M. G. C. (2004). Non-covalent interaction between procyanidins and apple cell wall material: Part I. Effect of some environmental parameters. *Biochimica et Biophysica Acta - General Subjects, 1672*(3), 192–202.

Le Bourvellec, C., & Renard, C. M. G. C. (2012). Interactions between polyphenols and macromolecules: Quantification methods and mechanisms. *Critical Reviews in Food Science and Nutrition, 52*(3), 213–248.

Lin, Z., Fischer, J., & Wicker, L. (2016). Intermolecular binding of blueberry pectin-rich fractions and anthocyanin. *Food Chemistry, 194*, 986–993.

Liu, X., Le Bourvellec, C., & Renard, C. M. G. C. (2020). Interactions between cell wall polysaccharides and polyphenols: Effect of molecular internal structure. *Comprehensive Reviews in Food Science and Food Safety, 19*(6), 3574–3617.

Malpezzi, L., Fronza, G., Fuganti, C., Mele, A., & Brückner, S. (2004). Crystal architecture and conformational properties of the inclusion complex, neohesperidin dihydrochalcone–cyclomaltoheptaose (β-cyclodextrin), by X-ray diffraction. *Carbohydrate Research, 339*(12), 2117–2125.

Mazza, G. & Kay, C. D. (2007). 10 Bioactivity, absorption and metabolism of anthocyanins. In: F. Daayf & V. Lattanzio, (Eds.). *Recent advances in polyphenols research.* (Vol. 1, pp. 117–126). Chichester, UK: Wiley Blackwell.

Nadar, S. S., & Rathod, V. K. (2017). Ultrasound assisted intensification of enzyme activity and its properties: A mini-review. *World Journal of Microbiology and Biotechnology, 33*(9), 170.

O'Donnell, C. P., Tiwari, B. K., Bourke, P., & Cullen, P. J. (2010). Effect of ultrasonic processing on food enzymes of industrial importance. *Trends in Food Science and Technology, 21*(7), 358–367.

Ortega-Regules, A., Romero-Cascales, I., Ros-García, J. M., López-Roca, J. M., & Gómez-Plaza, E. (2006). A first approach towards the relationship between grape skin cell-wall composition and anthocyanin extractability. *Analytica Chimica Acta*, *563*(1-2), 26–32.

Padayachee, A., Netzel, G., Netzel, M., Day, L., Zabaras, D., Mikkelsen, D., & Gidley, M. J. (2012). Binding of polyphenols to plant cell wall analogues - Part 1: Anthocyanins. *Food Chemistry*, *134*(1), 155–161.

Phan, A. D. T., D'Arcy, B. R., & Gidley, M. J. (2016). Polyphenol-cellulose interactions: Effects of pH, temperature and salt. *International Journal of Food Science & Technology*, *51*(1), 203–211.

Phan, A. D. T., Flanagan, B. M., D'Arcy, B. R., & Gidley, M. J. (2017). Binding selectivity of dietary polyphenols to different plant cell wall components: Quantification and mechanism. *Food Chemistry*, *233*, 216–227.

Plaschina, I. G., Braudo, E. E., & Tolstoguzov, V. B. (1978). Circular-dichroism studies of pectin solutions. *Carbohydrate Research*, *60*, 1–8.

Riou, V., Vernhet, A., Doco, T., & Moutounet, M. (2002). Aggregation of grape seed tannins in model wine - Effect of wine polysaccharides. *Food Hydrocolloids*, *16*(1), 17–23.

Trouillas, P., Sancho-García, J. C., De Freitas, V., Gierschner, J., Otyepka, M., & Dangles, O. (2016). Stabilizing and modulating color by copigmentation: Insights from theory and experiment. *Chemical Reviews*, *116*(9), 4937–4982.

Wang, D., Yan, L., Ma, X., Wang, W., Zou, M., Zhong, J., … Liu, D. (2018). Ultrasound promotes enzymatic reactions by acting on different targets: Enzymes, substrates and enzymatic reaction systems. *International Journal of Biological Macromolecules*, *119*, 453–461.

Wang, Y., Liu, J., Chen, F., & Zhao, G. (2013). Effects of molecular structure of polyphenols on their noncovalent interactions with oat β-glucan. *Journal of Agricultural and Food Chemistry*, *61*(19), 4533–4538.

Watrelot, A. A., Le Bourvellec, C., Imberty, A., & Renard, C. M. G. C. (2014). Neutral sugar side chains of pectins limit interactions with procyanidins. *Carbohydrate Polymers*, *99*, 527–536.

Wu, Z., Li, H., Ming, J., & Zhao, G. (2011). Optimization of adsorption of tea polyphenols into oat β-glucan using response surface methodology. *Journal of Agricultural and Food Chemistry*, *59*(1), 378–385.

Zhu, J., Zhang, D., Tang, H., & Zhao, G. (2018). Structure relationship of non–covalent interactions between phenolic acids and arabinan–rich pectic polysaccharides from rapeseed meal. *International Journal of Biological Macromolecules 66,* 144–153.

Acknowledgement/Danksagung

An dieser Stelle, möchte ich mich ganz herzlich bei den vielen besonderen Menschen bedanken, die mich während meiner Promotionszeit unterstützt, motiviert und begleitet haben.

Zunächst bedanke ich mich bei Herrn Prof. Dr. Schieber für die Möglichkeit und Betreuung der Promotion, das entgegengesetzte Vertrauen und die lehrreiche Zusammenarbeit während meiner Promotionszeit.

Des Weiteren danke ich Herrn Prof. Dr. Wüst für die Übernahme des Korreferats sowie Herrn Prof. Dr. Lamprecht und Frau Prof. Dr. Weisz für die Mitwirkung in der Promotionskommission.

Ein ganz besonderer Dank gilt Herrn PD. Dr. Fabian Weber für die intensive Betreuung und Motivation während der Promotionszeit, die Beiträge zu den Publikationen sowie für die Korrektur der vorliegenden Dissertation.

Ein besonderer Dank gilt ebenfalls Judith van der Weem für ihr großes Engagement und ihren Beitrag im Rahmen ihrer Masterabeit zur Unterstützung der vorliegenden Arbeit.

Ich bedanke mich darüber hinaus bei allen KollegInnen für die unvergessliche und familiäre Zeit im Arbeitskreis, aus der sich viele Freundschaften entwickelt haben. Ein Besonderer Dank gilt Rita Caspers-Weiffenbach und Timo Heinrichs für Ihre herzliche Art und die besondere Unterstützung im Labor. Lara Etzbach danke ich für die intensive Zusammenarbeit im letzten Jahr – eine bessere Partnerin kann man sich nicht wünschen.

Ein besonderer Dank gebührt meinen Eltern und meiner Schwester Jule, die mich während meiner gesamten Ausbildung unterstützt haben.

Abschließend gilt der wohl größte Dank meinem Freund Sebastian, der mich während der gesamten Zeit durch die Höhen und Tiefen begleitet hat und mir stets den Rücken gestärkt hat. Wir beide wissen nun, was folgendes Zitat bedeutet:

Success consists of going from failure to failure without loss of enthusiasm.

Winston Churchill

Curriculum Vitae

Lena Rebecca Larsen

Born on 16 October 1990 in Achim, Germany

Occupation

05/2016 – today	Rheinische Friedrich-Wilhelms University of Bonn Institute of Nutritional and Food Sciences, Molecular Food Technology

Academic studies

06/2016 – today	Rheinische Friedrich-Wilhelms University Bonn PhD studies in Food Chemistry
11/2015 – 04/2016	UCC University College of Cork, Ireland Research internship
10/2010 – 05/2015	Martin Luther University Halle-Wittenberg Diploma and 1st state examination in Food Chemistry
11/2014 – 05/2015	Fraunhofer Institute for Cell Therapy and Immunology IZI Diploma thesis: Characterization of the epitopes of β-conglycinin and profilin in the context of allergy research using phage display

School

08/2001 – 04/2010	Eichenschule Gymnasium Scheeßel

www.ingramcontent.com/pod-product-compliance
Ingram Content Group UK Ltd.
Pitfield, Milton Keynes, MK11 3LW, UK
UKHW022000190726
13853UKWH00004B/1651